Bridge Traffic Loading

Bridge Traffic Loading

From Research to Practice

Edited by
Eugene OBrien, Andrzej Nowak and
Colin Caprani

CRC Press
Taylor & Francis Group
Boca Raton London New York

CRC Press is an imprint of the
Taylor & Francis Group, an **informa** business

First edition published 2022
by CRC Press
2 Park Square, Milton Park, Abingdon, Oxon, OX14 4RN

and by CRC Press
6000 Broken Sound Parkway NW, Suite 300, Boca Raton, FL 33487-2742

CRC Press is an imprint of Informa UK Limited

British Library Cataloguing-in-Publication Data
A catalogue record for this book is available from the British Library

Library of Congress Cataloging-in-Publication Data
Names: O'Brien, Eugene J., 1958- editor. | Nowak, Andrzej S., editor. | Caprani, Colin, editor.
Title: Bridge traffic loading : from research to practice / edited by Eugene O'Brien, Andrzej Nowak, Colin Caprani.
Description: First edition. | Boca Raton : CRC Press, 2022. | Includes bibliographical references and index.
Identifiers: LCCN 2021020715 (print) | LCCN 2021020716 (ebook) | ISBN 9780367332464 (hbk) | ISBN 9781032101361 (pbk) | ISBN 9780429318849 (ebk)
Subjects: LCSH: Bridges--Live loads. | Load factor design.
Classification: LCC TG265 .B7925 2022 (print) | LCC TG265 (ebook) | DDC 624.2/5--dc23
LC record available at https://lccn.loc.gov/2021020715
LC ebook record available at https://lccn.loc.gov/2021020716

ISBN: 978-0-367-33246-4 (hbk)
ISBN: 978-1-032-10136-1 (pbk)
ISBN: 978-0-429-31884-9 (ebk)

DOI: 10.1201/9780429318849

Typeset in Sabon
by Deanta Global Publishing Services, Chennai, India

Contents

5 Long-span bridge loading   111

COLIN CAPRANI, MICHAEL QUILLIGAN, AND XIN RUAN

Preface

Bridges are subject to many of the same loads as other structures, such as self-weight, wind, earthquake, and differential settlement of foundations. What makes bridges different is that they are also subject to traffic – the weights of passing vehicles and the resulting dynamic interactions. Traffic loading is hugely uncertain – there are many different types of vehicle out there, varying most significantly in their weights, numbers of axles, and axle spacings. Of course, imposed loading in buildings is also quite varied and uncertain. What makes bridges different is that there are now big databases of truck weight data available. This has come from Weigh-in-Motion (WIM) systems that have been providing vehicle-by-vehicle weight data for many years now. Access to WIM data has led to statistical studies and research, much of it by the authors. As a result, it is now possible to calculate characteristic extreme load effects – moments, shears, etc. – with good accuracy.

Like other structures, bridges are designed to resist notional load models, deemed to reproduce extreme load effects, with adequate margins of safety. This book describes methods by which the characteristic extreme load effects can be calculated directly from the WIM data. This is essential information for those charged with developing notional load models for their networks. It is also extremely valuable for those situations where an existing bridge no longer has the capacity to carry the notional load model. Notional load models are necessarily conservative and it is often found that the characteristic extreme load effects are considerably less than what the notional load model generates.

We have targeted this book at a wide audience. Concepts familiar to those working in the field are explained pretty simply, with a view to bringing new entrants in the field up to speed. For example, probability paper plots are everyday tools in bridge traffic loading studies but we have explained them from first principles here. These are methods that have now reached such a level of maturity that consistent and reliable results can be derived from WIM data.

We have also gone well beyond established theory, introducing studies into the latest challenges such as traffic growth, vehicle/bridge dynamic interaction, and traffic microsimulation for long-span bridges. There are many challenges in the field, such as loading due to vehicles in multiple same-direction lanes and traffic that combines permit with non-permit vehicles. Not all of these challenges have been fully resolved but we have strived to present the current state-of-the-art, as it stands at the time of writing.

We have brought together experts from all over the world to deliver not just the state-of-the-art but also to provide guidance on how to address the issues. This book recommends best-practice for all the major challenges in the field – short-span, single- and multi-lane bridge loading, dynamic allowance, and long-span bridges. It also reviews some issues that continue to be debated – such as which statistical distribution is most appropriate, establishing when free-flowing or congested traffic governs, and how to deal with future traffic growth. Our intention has been to provide a 'one-stop-shop' on the topic of bridge traffic loading. We hope that you find it useful.

Acknowledgements

In addition to the co-authors of each chapter, the editors would like to recognize the contributions of several people who gave their time and expertise to assist us with specific parts of the text, or otherwise provided essential support. We thank the many researchers who have worked on many aspects of this topic – Dr. Daniel Cantero, Dr. Samuel Grave, Dr. Alisa Hayrapetova, Dr. Olga Iatsko, Dr. Alessandro Lipari, Dr. Alexandra Micu, Prof. Alan O'Connor, and Dr. Paraic Rattigan. We thank Dr Mayer Melhem, Monash University, for describing the Australian Standard's DLA in Chapter 4. We appreciate the willingness of FHWA and others to share Weigh-in-Motion data, particularly Rijkswaterstaat, the Dutch Ministry of Transport & Infrastructure, from whose WIM system, Figure 6.12, comes. We especially thank Dr Duy Do, Monash University, who managed to convert our sometimes badly-expressed ideas for figures into wonderful charts and illustrations throughout the book. Finally, we would like to thank our families for putting up with us while we disappeared at strange hours of the day and night to meet and discuss bridge loading, as we coordinated the writing of the book remotely across three time zones covering most of the globe.

Editors

Eugene OBrien is Professor of Civil Engineering at University College Dublin. He has previously worked in industry and led the study that resulted in an increase in the allowable weights of six-axle trucks in Ireland. He has worked on most aspects of bridge traffic loading including Weigh-in-Motion, loading on secondary roads, and loading on long-span bridges. He has pioneered new concepts such as scenario modeling, apparent permit vehicles, and micro-simulation modeling for long-span bridge load calculation.

Andrzej Nowak is Professor and Department Chair of Civil Engineering at Auburn University. His development of a reliability-based calibration procedure for calculation of load and resistance factors has been successfully applied to calibration of the AASHTO design code for bridges, ACI 318 Code for Concrete Buildings and Canadian Highway Bridge Design Code. He has an Honorary Doctoral Degree from Warsaw University of Technology, he is a Fellow of ASCE, ACI, and IABSE, and he has received the ASCE Moisseiff Award, IFIP WG 7.5 Award, Bene Merentibus Medal, and the Kasimir Gzowski Medal from the Canadian Society of Civil Engineers.

Colin Caprani is Associate Professor in Structural Engineering at Monash University and a Chartered Structural Engineer. He has made fundamental contributions to highway bridge trafficloading, and his work focuses on the safety assessment of existingstructures leveraging structural health monitoring and reliabilitytheory. He has applied his research to practice, enabling hundreds ofsuperload transport movements, and developed the Austroads guidelineon probabilistic bridge assessment. He is a code committee member forAS 5100 Part 2 – Loads, Part 5 – Concrete, and Part 7 –Assessment, and an Associate Editor of the *ASCE Journal of Bridge Engineering* and *Structural Safety*.

Contributors

Jacek Chmielewski
Department of Civil Engineering
Bydgoszcz University of Science and
Technology
Bydgoszcz, Poland

Bernard Enright
Department of Civil and
Environmental Engineering
Technological University Dublin
Dublin, Ireland

Donya Hajializadeh
Department of Civil and
Environmental Engineering
University of Surrey
Guildford, UK

Jan Kalin
Slovenian National Building and
Civil Engineering Institute
Ljubljana, Slovenia

Jennifer Keenahan
School of Civil Engineering
University College Dublin
Dublin, Ireland

Cathal Leahy
Advanced Digital Engineering
Arup
Dublin, Ireland

Roman Lenner
Deptartment Of Civil Engineering
Stellenbosch University
Stellenbosch, South Africa

Michael Quilligan
School of Engineering
University of Limerick
Limerick, Ireland

Xin Ruan
Department of Bridge Engineering
Tongji University
Shanghai, China

Sylwia Stawska
Department of Civil and
Environmental Engineering
Auburn University
Auburn, Alabama

Ales Znidaric
Slovenian National Building and
Civil Engineering Institute
Ljubljana, Slovenia

Chapter 1

Introduction

*Andrzej Nowak, Sylwia Stawska, and
Jacek Chmielewski*

1.1 INTRODUCTORY REMARKS

Highway bridges are subjected to various loads, including dead load, traffic load, the influence of temperature, wind, earth, and water pressure, and accidental forces such as collision and fire. Bridge owners are interested in the safety of the structure, which is equally affected by the applied loads and its capacity to resist them. Imposed load is an important load component, representing the effect of moving traffic. This book is focused on traffic load, derived from traffic volume, the weights of vehicles, and their axle configurations. Load effects (LEs), such as moments and shear forces, are found and the data is extrapolated to find characteristic maximum values. The presentation is at a general level and is supplemented with data and examples.

There is a wide variety of vehicles that differ with regard to the number of axles, axle spacing, and axle load. The variation can depend on geographical region, country, and/or other local characteristics. The objective of the book is to review the body of knowledge related to traffic loads and serve as a guide for bridge engineers and engineering students. There is an extensive database available from the Weigh-in-Motion (WIM) stations located in the United States, Europe, Australia, and others. This database can be used to verify design traffic load provisions and to customize these for local site conditions. Transportation structures such as roads and bridges are designed to carry moving traffic loads. However, excessive static and dynamic traffic LEs can cause damage or even collapse of structural components or whole structures. Many bridges are approaching the end of their design lives (e.g. ASCE 2017). A better knowledge of the applied load on these bridges facilitates a more accurate safety rating and hence a better prioritization of bridges for repair or replacement.

Highway traffic can be considered as a mix of three categories of vehicle: (1) legal vehicles, (2) permit vehicles, and (3) illegally overloaded vehicles. Illegally overloaded vehicles are a source of potential damage to bridges and increase the risk of failure. Permit vehicles are often transporting large

DOI: 10.1201/9780429318849-1

1

indivisible cargo (e.g. cranes, agricultural or military machinery) and are sometimes subject to speed and other restrictions.

The problem of unrestricted operation of short trucks in the United States with GVW below 35 tonnes is considered by Sivakumar (2007) using a large WIM database collected from 18 states during 2001–2003. Special vehicles that do not meet the US federal bridge formula requirements were the subject of a study to develop notional legal trucks for the AASHTO bridge rating policy. The consequences of permit violators for the infrastructure are estimated by Luskin & Walton (2001), considering different regulatory scenarios. The benefits of filtering WIM data to separate permit vehicles from illegal ones are shown by Caprani et al. (2008). In Wisconsin, individual vehicle records are used to evaluate the state-specific standard permit vehicles based on a statistical analysis of the LEs caused by the heaviest 5% of trucks in each vehicle class (Zhao & Tabatabai 2012). Similarly, both European and US WIM databases are used to identify probable-permit vehicles and to produce equivalent permit truck traffic using Monte-Carlo simulation by Enright et al. (2016). Fiorillo & Ghosn (2014) propose a sorting procedure to define the proportions of illegally overloaded and permitted traffic, based on WIM data collected by the New York Department of Transportation. However, there is still no exact method available to distinguish permit vehicles from illegally overloaded ones in a WIM dataset (Enright et al. 2016).

Moving vehicles result in additional dynamic forces. The estimation of the actual dynamic load is based on the results of field measurements. For comparison, the dynamic design loads are presented for selected bridge design codes in the United States, Europe, and Australia. Traffic load on long-span bridges is a result of multiple presence of vehicles, typically a long line of mixed congested traffic resulting from an accident.

The aim of this book is to present a compilation of the available knowledge related to road traffic load on bridges, its background, current trends, and prediction of future changes. It provides a review of the history of traffic loads and the evolution of design-imposed load in the design codes. It reviews the information that can be deduced about current traffic load from an analysis of WIM measurements. Special consideration is given to dynamic vehicle/bridge interaction, load on long-span bridges, and prediction/extrapolation of future extreme traffic loading. It is intended for bridge and traffic engineers, researchers interested in bridge engineering, faculty teaching bridge and transportation-related courses, and students.

1.2 DESIGN TRAFFIC LOAD

The role of design codes or standards is to allow bridges to be designed while ensuring adequate safety. The design loading on bridges is therefore a notional loading, deemed to represent the worst combination of vehicles, i.e. the characteristic maximum, in a specified return period. The return period ranges widely, from 75 years in the United States to 1000 years in Europe and should not be confused with the design life. The return period is a probability level, e.g. the level of loading expected to be exceeded just once in 1000 years. The design life, e.g. 100 years, is the intended working life of the bridge. Coincidentally, the same value of 75 years is specified for the return period and the design life in the AASHTO standard.

In the current generation of codes, safety is provided through load and resistance factors which are applied to the notional loading and the material properties. It is not appropriate to compare levels of safety between codes or countries without considering the complete picture of load, resistance, and their factors. The design traffic load typically evolves over the years to keep up with a general international trend towards heavier and more frequently heavy vehicles. The historical evolution of design traffic load for bridges is shown using the example of the AASHTO specifications. The design code must cover the broader perspective of bridge design, where not only traffic volume, weight, and configuration are taken into account, but also lane loading, the possibility of multiple presence of heavy vehicles on the bridge, and dynamic vehicle/bridge interaction.

In the United States, bridges were designed in the 1930s for notional load combinations known as H-10 and H-15 (see Figure 1.1). In 1944, the HS-20 was introduced, and remained the basis for bridge design until 2007. The HS-20 truck is considered to be a conservative representation of a 5-axle vehicle, with the 142 kN (32-kip) axles representing the effects of tandem axle groups. Three different forms of the HS-20 loading were specified with the requirement that the most critical be considered for the calculation of LE.

In the 1970s it became clear that the HS-20 load model was no longer an adequate representation of the extremes of existing traffic-induced loading. In Canada, a new live load was specified in the first edition of the Ontario Highway Bridge Design code (OHBD) in 1979. This was the first North American design code based on the philosophy of limit states. The second edition of the OHBD code was published in late 1983. The OHBD live load was specified as the OHBD Truck shown in Figure 1.2 or 70% of the OHBD Truck loads superimposed on a 3.0 m wide uniformly distributed load of 10 kN/m.

In 1987 AASHTO initiated a feasibility study and in 1994, the AASHTO Load and Resistance Factor Design (LRFD) specifications introduced the HL-93 model, which has been the basis for design of bridges in the United States since 2007. In AASHTO LRFD, three different scenarios of HL-93 loading are specified, as shown in Figure 1.3. The development of HL-93 was carried out as part of National Cooperative Highway Research Program (NCHRP) Project 12-33. The calibration procedure and results are documented in NCHRP Report 368 (Nowak 1999). The traffic load model is based on a traffic survey from Ontario (Canada) that consisted of 9250 selected trucks since US data was not available at that time

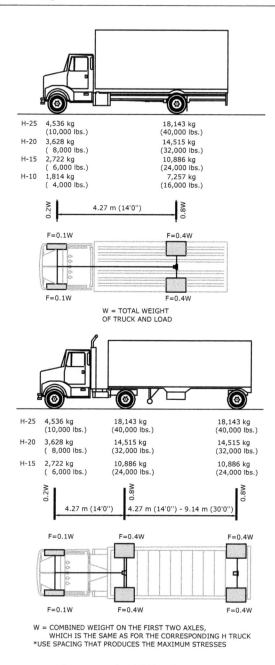

Figure 1.1 Design truck configurations for AASHTO standard specifications for highway bridges (W is GVW).

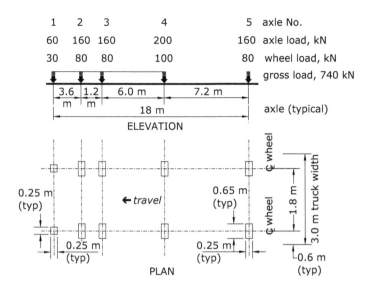

Figure 1.2 OHBD Truck (Ontario Highway Bridge Design Code 1979).

(Agarwal & Wolkowicz 1976). Since then, an extensive WIM database has been established by State DOTs, the Federal Highway Administration (FHWA), and as part of other NCHRP studies (Allen et al. 2005, Ghosn et al. 2010, Kwon et al. 2011, Sivakumar et al. 2011, SHRP2 2015, Nowak & Iatsko 2017).

The design live load for continuous spans allows two trucks to be applied, one in each of the adjacent spans, superimposed with a uniformly distributed load of 9.3 kN/m. However, because of the reduced probability of a simultaneous occurrence of very heavy trucks, the resulting negative moment is reduced by 10%.

In AASHTO Load Factor Design LFD, a 'traffic load factor' is specified of 1.3 (5/3) = 2.17, where 1.3 is a load factor applied to all loads and 5/3 is the live load factor. In AASHTO LRFD, the live load factor is 1.75. A comparison of unfactored traffic load moments and shear forces in simply supported beams is shown as a function of the span in Figure 1.4. It can be seen that HL-93 is considerably more onerous for moment, particularly for longer spans. On the other hand, it is less onerous than the older HS-20 model for shear force.

In AASHTO LRFD, dynamic load is specified as an additional 33% of live load but applied to the design truck only (no dynamic load applied to the uniformly distributed lane load).

AASHTO LRFD specifies 3.66 m (12 feet) wide notional lanes, rounded down to the nearest integer value. Depending on the number of loaded traffic lanes, a multilane reduction factor is applied. It takes the two-lane

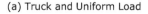

(a) Truck and Uniform Load

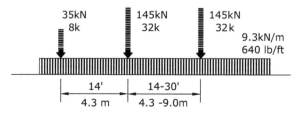

(b) Tandem and Uniform Load

(c) Alternative Load for Negative Moment (reduce to 90%)

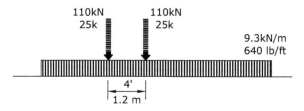

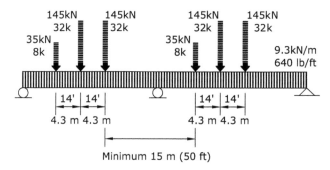

Figure 1.3 HL-93 loading scenarios in AASHTO LRFD specifications.

bridge as the benchmark, so the reduction only applies for three lanes or more and the factor increases the load on single-lane bridges. Thus, the reduction factor is 1.2 for one loaded lane, 1.0 for two, 0.85 for three, and 0.65 for four loaded traffic lanes.

The Eurocode (CEN 2003) specifies a 'Normal' load model, deemed to represent the extremes of regular (non-permit) vehicle traffic and to include an allowance for dynamics. A list of possible 'Abnormal' (permit) vehicle configurations is also presented, and national standards authorities are free to specify which abnormal vehicles should apply for each class of road. The Normal load model, shown in Figure 1.5, is intended to generate the characteristic maximum LEs corresponding to a 5% probability of exceedance in 50 years. This corresponds to a return period of about (50/0.05 =) 1000 years.

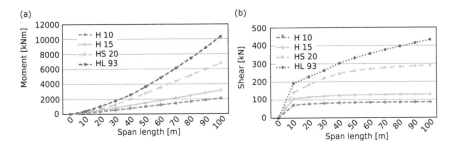

Figure 1.4 Comparison of LEs in simply supported beams due to historic and current AASHTO load models: (a) moment; (b) shear force.

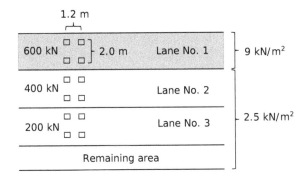

Figure 1.5 Eurocode Normal road traffic load model.

Like the AASHTO model, the Eurocode load model, shown in Figure 1.5, is applied to notional lanes that are independent of the actual lanes marked on the road. In this case, each notional lane is 3 m wide. Any remaining road width between curbs is referred to as the remaining area. The notional load model consists of uniform loading and a tandem in each lane with four wheels. The uniform loading is 9 kN/m² in the lane with the most severe effect and 2.5 kN/m² everywhere else, including the remaining area. The tandems weigh 600 kN, 400 kN, and 200 kN (combined weight of all four wheels in each case) and are placed in the most adverse locations. All of these loads include the effect of dynamic vehicle/bridge interaction (see Chapter 4).

Moving trucks create other load effects that have to be considered in the design of bridges. These effects include centrifugal forces, and forces caused by braking and collision. Vehicles traveling on curved segments of road create forces that have a component that is perpendicular to the direction of travel. The magnitude of these centrifugal forces depends on the vehicle speed and curvature of the road. In the AASHTO standard, it is assumed that the centrifugal force is applied at 1.8 m (6 ft) above the road surface which corresponds to the gravity center of the vehicle. Braking causes a horizontal force in the direction of traffic and it is also assumed in AASHTO to be applied 1.8 m (6 ft) above the wearing surface. Bridge structures can be affected by other traffic related loads, in particular vehicle collision forces. The structure has to be either designed for an impact load or protected by barriers or other impact absorbing structures.

1.3 STATISTICAL BACKGROUND TO BRIDGE LOADING

Notional load models are intended to represent the characteristic maximum LEs, i.e. those levels of LE that would be expected to be exceeded just once in the specified return period. Block maximum values, e.g. maximum-per-day LEs, are a convenient way of quantifying the problem. In these calculations, weekends and public holidays tend to be discarded as there are fewer heavy vehicles, and there are assumed to be 250 working days per year (365 – 52×2 weekend days – 11 public holidays). Then, for example, a 75-year return period corresponds to a probability level of 1 in 75×250 or 1 in 18,750. The probability of non-exceedance is then (1 – 1/18,750 =) 0.999947. The cumulative distribution function of maximum-per-day data is a plot of cumulative probability against LE (Figure 1.6). Hence, the characteristic maximum value can be identified on this graph as the value corresponding to the probability of non-exceedance, i.e. 0.999947 in this example.

The cumulative probabilities in this graph are generally rescaled to illustrate the extremely rare data better. These rescaled cumulative distribution graphs are known as probability paper plots. A commonly used scale is the inverse cumulative distribution function for a Normal distribution,

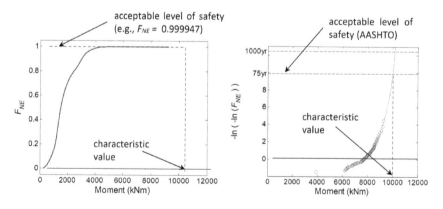

Figure 1.6 Maximum-per-day LE data for mid-span bending moment in a 40 m long bridge: (a) cumulative distribution function; (b) probability paper plot.

and data plotted to such a scale will appear as a straight line if it is from a Normal distribution. The construction and use of Normal probability paper is described in textbooks on probability and statistics, e.g. Nowak & Collins (2013).

Block maximum data are the maximum of many other values – for example, maximum-per-day LE is the maximum of all LE values that occur in a day. There are three statistical distributions used for this kind of extreme value data: Gumbel, Weibull, and Fréchet, also known as Extreme Value distributions, Type 1, 2, and 3. Extreme value data is often plotted on a double-log or Gumbel scale, as double-log is the inverse cumulative distribution function for the Gumbel distribution. When plotted to a Gumbel scale, Gumbel, Weibull, and Fréchet data appear as straight line, concave (curving upwards), and convex, respectively. Figure 1.6 (b) is a probability paper plot of the same data as Figure 1.6 (a). The non-exceedance probability values corresponding to 75 year and 1000-year return periods are shown in this graph for reference (AASHTO and Eurocode).

The data in Figure 1.6 (b) is clearly concave, suggesting that it is consistent with a Weibull distribution. It is often argued that vehicle weight per unit length has a physical upper limit, dictated by the load-carrying capacity of the vehicle suspensions. It follows that LE has a physical upper limit. This manifests itself in a Gumbel probability paper plot as data trending towards an asymptote, i.e. towards a limiting value, as the event becomes increasingly rare and the probability of non-exceedance tends towards unity.

Chapter 2

Vehicles and gross vehicle weight

Andrzej Nowak, Jacek Chmielewski, and Sylwia Stawska

Traffic flow is made up of vehicles that can be sorted into groups, depending on their axle spacings and loads. Vehicles can be considered as standard legal, standard illegal, or permit:

- Standard legal vehicles meet the general regulations for axle spacings and weights and do not require a permit, and special vehicles that have exemptions under grandfather provisions.
- Standard illegal vehicles do not meet the regulations of axle spacings, axle, and/or GVW.
- Permit vehicles are outside the general regulations for axle spacings and require a permit. They may or may not have this permit and may be in breach of it so they may be legal or illegal.

Eurocode 1 (CEN 2003) defines loading due to standard legal and illegal vehicles as 'normal' and loading due to permit vehicles as 'abnormal.' Bridge design load models such as HL-93 and the Eurocode 1 model, are intended to represent the extremes of normal loading, i.e. standard vehicles. Abnormal loading is considered separately – typically a permit is only issued when the bridges on the allowable routes have been rated as having sufficient capacity to carry the specified vehicle.

Countries have legal load limits defined, not only in terms of GVW, but also in terms of axle loads, axle group loads, number of axles, and axle spacings. Usually, a small number of vehicle types tend to govern for characteristic maximum bridge load effects. The critical load effects for each bridge strongly depend on the distribution of GVW but also on the traffic mix and the dominant vehicle types. From the perspective of bridge loading, the major factors in traffic data are volume, vehicle weights, axle configuration, and multiple presence, i.e. the simultaneous occurrence of multiple vehicles on the bridge, either within a lane or in adjacent lanes.

DOI: 10.1201/9780429318849-2

2.1 CATEGORIES OF VEHICLE

There are a great variety of vehicle classification systems around the world. In the US, vehicle classes are specified according to the Federal Highway Administration (FHWA) system (Cambridge Systematics 2007), as shown in Figure 2.1. Categories of vehicle depend on whether they carry passengers or freight. Non-passenger vehicles are further subdivided according to the number of axles and the number of tractor/trailer units. WIM databases typically provide information about the vehicle's class. Automatic classifiers use an algorithm to interpret axle spacing and total vehicle length information and infer from this the necessary information on the number of units. Vehicles are sometimes mis-classified but it should generally be a small percentage of the total. In most countries, vehicles up to 6 axles (up to Class 9 or 10) are considered standard and Classes 11–13 require permits.

Using the FHWA classification scheme, vehicle Classes 9 and 10 tend to dominate on the world's highways – see Figure 2.2. This is the 5- or 6-axle tractor/semi-trailer, used for long haul freight transport. It typically consists of a steering axle, a second tractor axle, or tandem, and a trailer tandem or tridem group. As many axles in modern trucks are liftable, details of the configuration can change as the driver lifts an axle.

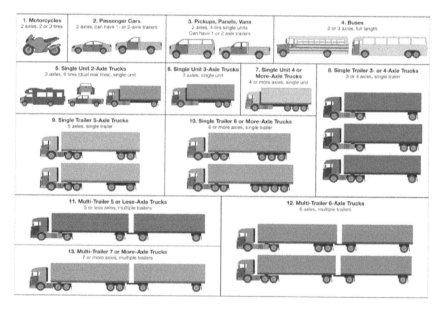

Figure 2.1 United States FHWA Vehicle Classification Scheme (Cambridge Systematics 2007).

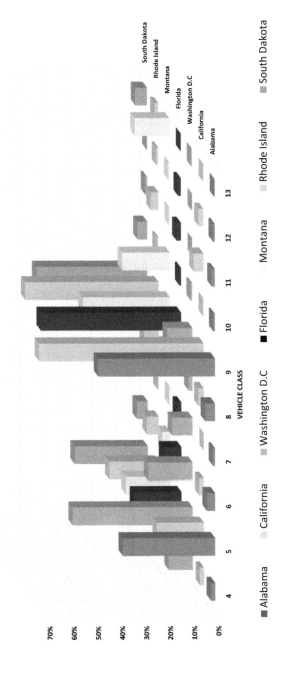

Figure 2.2 Vehicle class distributions at a number of US WIM sites (FHWA classification system).

2.2 TRUCK WEIGHT DATA

Vehicle weights are clearly fundamental to any consideration of bridge traffic loading. The most accurate systems for weighing trucks are static weigh stations. These are located off-road and typically have static scales built into the pavement. For enforcement systems, heavy vehicles are directed to exit the highway to be weighed – Figure 2.3. Some of these systems use low-speed WIM which is not fully static but where the speed is very low, and the transverse position is controlled by curbs. In these enforcement systems, an operator then checks if the legal weight limits have been violated. Static weigh stations measure only a tiny fraction of the traffic. Further, their locations are known by truck drivers and illegally overloaded vehicles may try to avoid them, resulting in biased truck weight data.

Portable axle weigh scales, Figure 2.4, can be set up in any large level area. However, the setup process is time consuming (about 45 minutes per truck) and labor intensive. If the weighing area is on-road, the process can cause traffic disruption and may increase the possibility of a traffic accident.

At present, the major source of information about the weights of vehicles is the network of WIM systems. While WIM is being used for direct enforcement of overload in some countries (Žnidarič 2015), the accuracy of most WIM systems is not sufficient for direct enforcement and they are generally only used for data collection. WIM systems have been collecting good quality data for more than 20 years and, at the time of writing, there are millions of vehicle records at some locations. WIM systems and Continuous Count Stations (CCSs) are the two primary sources of traffic data (Hallenbeck & Weinblatt 2004). CCSs are also referred to as Counter and Classifier sites or Automatic Traffic Recorders. A continuous count is a volume count derived from permanently installed counters. Counts can be done for 24 hours each day over 365 days per year. There are some different techniques used by CCSs, and the counting and vehicle classification

Figure 2.3 Weigh Station.

Figure 2.4 Portable scales weight measurements (Žnidarič 2015).

accuracy are subject to the CCS operation environment. Most of CCSs only collect traffic volume, vehicle class, and some of them vehicle speed data, whereas WIM systems collect this plus the load spectra, i.e. the histograms of axle and gross vehicle weights. WIM data is a powerful first enabler of traffic load assessment and facilitates the development of statistical models of bridge traffic load. Each traffic record includes a detailed description of the vehicle configuration, exact time and date, lane and direction code, speed, GVW, individual axle loads, axle spacings, and class of vehicle (Cambridge Systematics 2007).

One of the first WIM systems was developed in 1952 by the United States Bureau of Public Roads, which was a predecessor of the FHWA (Norman & Hopkins 1952). It was just a reinforced concrete platform instrumented with electrical resistance strain gauges. The vehicle weight was calculated manually using an oscilloscope attached to the strain gauges. Contemporary WIM systems are very different from the early sensors developed in the 1960s. In addition to axle loads, modern systems determine the vehicle type, and they process and transmit the recorded data (AASHTO 2014). Currently, there are over 700 WIM stations in operation in the United States and thousands worldwide (Ghosn et al. 2010).

There are several factors which can affect the accuracy of measurements collected by any type of WIM system, such as pavement roughness (causing bouncing axle movement or dynamic impact), temperature effects, and gradients in the road. ASTM E1318-09 (ASTM 2009) classifies WIM systems into four types: Type I through Type IV, depending on performance requirements, with Type IV expected to be the most accurate. The COST 323 draft specification (Jacob 1995) defines a more elaborate framework for the accuracy classification of WIM systems. In general, a Class A system, according to COST 323, has 95% of gross weights within an accuracy of ±5%, axle group weights (tandems and tridems) within 7%, and single axle weights within 8% . However, the confidence level of 95% is adjusted, depending on the test conditions (number of vehicles, season(s) when tested, whether repeated runs or general traffic).

There is a variety of Weigh-in-Motion technologies available for permanent or temporary traffic data collection. Sensor types include piezo-polymer and piezo-ceramic, piezo-quartz, bending plate, load cell, and Bridge WIM (Al-Qadi et al. 2016, McCall & Vodrazka 1997). Piezo-polymer and piezo-ceramic sensors come in strips that are placed in a groove in the pavement surface; see Figure 2.5. They are highly temperature sensitive and are mostly used for vehicle count and classification (Al-Qadi et al. 2016). Piezo-quartz sensors also come in strips but are made up of a series of discrete sensing elements. They have a lower sensitivity to temperature fluctuation (White et al. 2006) and belong to the ASTM E1318 Type I systems category. Piezoelectric sensors record a change in voltage in response to the change in pressure due to the passing wheel. They are therefore only effective for dynamic load and are not suitable for static or slow-speed measurements. Piezo-quartz sensors are also pressure sensitive but utilize a quartz crystal force sensing technology. They contain an aluminum alloy profile in the middle of which quartz discs are fitted every 5 cm. All of these strip sensors are installed flush with the asphalt or concrete pavement surface, generally with epoxy adhesive.

The bending plate sensor (Al-Qadi et al. 2016a, McCall & Vodrazka 1997), illustrated in Figure 2.6, consists of strain gauges attached to a steel plate that bends as wheels pass over it. They are generally insensitive to

Figure 2.5 WIM Strip Sensor.

Figure 2.6 Bending Plate WIM system.

temperature fluctuations. In load cell-based WIM systems, the reactions are measured in the supports of a large plate. The load cells at each support generally use strain gauges to determine the applied forces (Al-Qadi et al. 2016b) and generally provide good accuracy.

There are now numerous WIM stations all over the world, collecting millions of truck records. Researchers have used WIM data from these databases to develop more efficient bridge designs (Nowak 1999, Nowak & Iatsko 2017), to evaluate existing bridges (Nowak & Tharmabala 1988, Sivakumar 2007), and for fatigue studies (Fisher et al. 1983, Laman & Nowak 1996, Kulicki et al. 2015).

The changes in truck traffic volume, weight, and configuration in recent decades are reviewed by Anitori et al. (2017), Ghosn et al. (2010), Liao et al. (2015), Babu et al. (2019), and Treacy & Brühwiler (2012). A set of protocols and techniques for the collection and analysis of WIM data, along with methods for the calculation of traffic load factors to be used in American LRFD design, are presented by Ghosn et al. (2010).

Virtual Weigh Station (Figure 2.7) is a term used to describe a WIM scale used as an enforcement facility, along with digital cameras and software to process the information in real-time. It does not require continuous staffing and typically it is monitored from another location.

A study in Alabama (Stawska et al. 2021) is used here to illustrate some typical WIM system accuracies. The accuracy of WIM data can be measured in different ways. There are four vehicle parameters of importance: (i) GVW, (ii) group axle weights, (iii) individual axle weights, and (iv) weights of axles within a group – see Figure 2.8. The definition of an axle group varies but is usually a function of the axle spacing – in the Alabama study, a tandem group is defined as two axles less than 1.8 m (6 feet) apart. Axle spacings are also important for studies of bridge loading and the vehicle time stamps, which need to be accurate to 0.01 s to facilitate calculations of inter-vehicle distances.

Figure 2.7 Virtual WIM Station (IRD 2020).

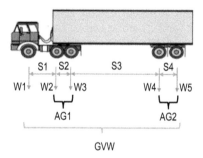

Figure 2.8 Parameters of a typical vehicle.

In the Alabama study example, the accuracy of measurements was checked by a statistical analysis of vehicle parameters using data collected using four different weighing techniques, weigh station, portable scales, WIM, and bridge WIM (B-WIM). The weigh station measurements are taken here as the reference point. The linear correlation coefficient between weigh station and other weighing techniques is presented in Figure 2.9. For B-WIM the results indicate the positive correlation coefficients, varying between 0.64 to 0.81. Lower values of correlation coefficient are noted for the portable scale data – they vary from 0.59 to 0.77 (moderate/high positive correlation). The lowest values of correlation coefficient are obtained

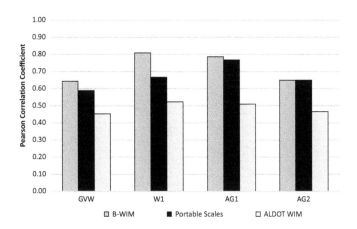

Figure 2.9 Correlation coefficient between measured values recorded by Weigh Station and other measuring techniques in Alabama.

from the WIM data – these are in the range from 0.45 to 0.52 (moderate positive correlation).

For dynamic weighing measurements, it is required to verify the data accuracy, and eliminate questionable records, before using it to assess the traffic-induced load effects on bridges. Tolerance checks were carried out on the Alabama data, according to the ASTM E1318-09 (ASTM 2009) requirements. For a Type I WIM system, these are that 95% of GVW, axle group, and individual axle results should be within ±10%, ±15%, and ±20% respectively. The WIM system accuracies for GVW are illustrated in Figure 2.10. The static weigh station weights are used as the reference so results from this system will fall on the diagonal (45°) line. Lines are also shown representing results falling ±10% from the static system. The static portable system is mostly in a range of ±10% of weigh station measurement, but WIM and B-WIM do not meet these tolerances.

2.3 QUALITY CONTROL OF TRAFFIC DATA

The major source of information about traffic loading is the WIM database, and poor-quality data can lead to poor quality bridge load effect calculations. Data errors can result from many issues, including WIM system malfunction, loss of calibration, poor temperature compensation, and vehicle miss-classification. Quality Control is an important process of removing suspect data and improving overall data quality.

Two types of error can occur in long-term WIM data collection: random errors (affecting individual vehicles) and systematic errors (occurring frequently and affecting groups of records). There are case studies related

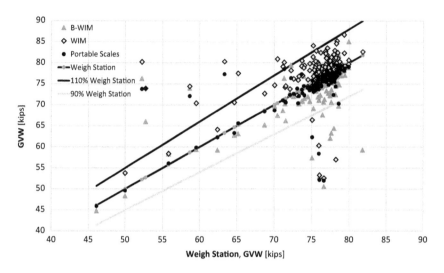

Figure 2.10 GVW accuracy of WIM technologies in Alabama test.

to quality checks of traffic data that are adopted by many state agencies in the US (Babu et al. 2019, Elkins & Higgins 2008, Southgate 1990, Ramachandran et al. 2011, Qu et al. 1997, Quinley 2010, Kulicki et al. 2015). However, there is no universal documented Quality Control procedure. All states gather traffic data as part of FHWA's Highway Policy Management System and the quality has to meet the minimum requirements prescribed in the guides (Quinley 2010). Some US states have developed their own Quality Control programs to meet customer needs and achieve maximum performance (Vandervalk-Ostrander 2009). The need for adequate quality of traffic data in bridge design has been studied extensively in (Ghosn et al. 2010, Sivakumar et al. 2007).

Important documents that provide guidelines for WIM data Quality Control are the Traffic Monitoring Guide, AASHTO Guidelines for Traffic Data Programs, and the Highway Performance Monitoring System Field Manual (HPMS) (Vandervalk-Ostrander 2009). In addition, the Long-Term Pavement Performance program collects traffic data as a part of the pavement study (Elkins et al. 2018) and specifies Quality Control procedures for WIM data (Walker & Cebon 2012, Walker et al. 2012). More literature related to Quality Control checks in national standards and common practice are discussed in the following sections.

The collected traffic data are recorded using different formats. For instance, in the Traffic Monitoring Guide there is a Station Description format, Traffic Volume format, Vehicle Classification format, Weight format, and five other formats (FHWA 2016). In the Long-Term Pavement Performance system, different formats can be used depending on the type of software which is used to process the WIM data. The Long-Term Pavement Performance Traffic Quality Control software has 4-card (Classification card) and 7-card (Weight card) data formats. At many WIM locations, the data is processed by vendor's software that can produce data in a variety of formats. The Traffic Monitoring Guide contains a compendium of Quality Control criteria used by various states and recommends the checks used in the Traffic Monitoring Analysis System. Before data is updated, it is filtered through a number of checks. The AASHTO Traffic Data Program guidelines recommend minimum validation criteria for weight, classification, and vehicle count data. The Long-Term Pavement Performance database has the most rigorous Quality Control checks. Traffic data stored in this database has to comply with the Quality Control checks mentioned in its Information Management System Manual (Elkins et al. 2018, FHWA 2016).

In the US, WIM data can be utilized for the evaluation of existing bridges (AASHTO 2018), but the procedure to assess the quality of the WIM data is not discussed in detail. Many filtering criteria have been developed in previous studies to improve the quality of traffic data. Characteristic maximum traffic loading on bridges is determined by an extrapolation process (Nowak 1993, Iatsko 2018). The threshold limits that are used to filter the data may impact the upper tail of the distribution of traffic data that is used in bridge traffic load modeling. Clearly these tail data – the biggest and heaviest vehicles – are highly significant in any assessment of traffic load and their accuracy is key. The importance of the upper tail is discussed more by OBrien et al. (2010).

An example of a Quality Control procedure is proposed by Babu et al. (2019) to process and eliminate erroneous records. Firstly, identical rows are sought, which can be an indication of a typical system malfunction. Then, the Quality Control procedure seeks to find errors in the description, vehicle configuration, vehicle weight, speed, and other more advanced checks. The proposed Quality Control checks are shown in Table 2.1.

An alternative filtering scheme, used to clean European WIM data is presented by Enright & OBrien (2011). An investigation of inter-vehicle (axle to axle) gaps identified several cases where gaps were below 0.2 s. Photographs confirmed these to be the result of trailers being mis-classified as separate vehicles. A comparison of lengths found vehicles where wheelbase (first-axle-to-last-axle length) exceeded total length. This was the result of 'ghost' axles – the software typically replicated the rear tandem, making 5-axle trucks into 8-axle vehicles. Based on European WIM data processing experiences, a set of scores was proposed to identify doubtful records. This allows vehicles with one slightly doubtful attribute to be retained but records with multiple doubtful attributes to be rejected. The set of rules, along with scores, are presented in Table 2.2. A score of 7 results in rejection of the record.

Table 2.1 Quality Control Filtering Criteria (Babu et al., 2019)

Type	Filtering criteria	Threshold limits
WIM description	Station ID	Null or invalid state ID
	Lane of travel	$\neq$ (0–9)
	Direction of travel	$\neq$ (0–9)
Time stamp	Invalid year	Null or irrespective year
	Invalid month	$\neq$ (1–12)
	Invalid day	$\neq$ (1–31)
	Invalid time	$\neq$ (0–86399) sec.
Duplicates	Identical records	Exact copy
	Same axle weight for consecutive axles	Axle weight = Axle weight n+1= …
Vehicle configuration	Invalid vehicle class	$\neq$(1–13)
	Zero GVW	= 0
	Zero axle spacings	= 0
	Number of axles	$\neq$ (2–22)
	Number of axle weights	$\neq$ (2–22)
	Number of axles is equal number of recorded axle weights	Number of axles = Number of axle weights
	Number of axles spacings	$\neq$ (1–21)
	Number of axles is equal to number of axles spacings + 1	Number of axles $\neq$number of axles spacing +1
	Sum of axle weights is equal GVW $\pm$ 10%	$\pm$ 10% of GVW
	Minimum first axle spacing	< 6 ft
	Minimum axle spacing	< 3.3 ft
	Steering axle weight	> 40 kips
	Single axle weight	$\neq$ (1–60 kips)
	Tandem axle weight	> 60 kips
	Tridem axle weight	> 80 kips
	Average left and right wheel weight	> $\pm$20%
	Total vehicle length	> 220 ft.
Speed limits	Vehicle speed	$\neq$ (10–90 mph)

2.4 VARIATIONS IN WIM DATA

In this section, differences in traffic parameters between the United States and Europe are presented. The traffic is compared for several states in the US and selected European countries. The traffic is very site specific which makes it challenging to develop a consistent national traffic load model for bridge design and evaluation. The WIM data vehicles are compared in terms of GVW, and tandem and tridem axle weight.

Table 2.2 WIM Data Attribute Check with Scores (Enright & OBrien 2011)

Attribute	Points
Rules applied to overall vehicle:	
GVW less than 3.5 t (cars)	7
Wheelbase less than 1 m	7
Wheelbase greater than 30 m and first or last axle spacing greater than 10 m	7
Wheelbase greater than 30 m and speed less than 30 km/h	7
Wheelbase greater than 40 m	7
Maximum axle load greater than 15 t and this axle represents more than 85% of GVW	7
Speed less than 20 km/h	7
Speed greater than 120 km/h	7
Speed between 20 and 40 km/h	+5
First axle spacing greater than 15 m	7
First axle spacing greater than 10 m	+4
Rules applied for each axle:	
Any left or right wheel weight zero or negative	7
Ratio of left/right wheel weights > 5	7
Any axle load zero or negative	7
Axle load greater than 60 t	7
Axle spacing greater than 20 m	7
Points accumulated per axle:	
Ratio of left/right wheel weights between 2 and 3	+1
Ratio of left/right wheel weights between 3 and 5	+2
Axle load between 25 t and 40 t	+2
Axle load between 40 t and 60 t	+5
Axle spacing less than 0.4 m	7
Axle spacing between 0.4 and 0.7 m	+2
Axle spacing between 0.7 and 1.0 m	+1

Differences between European and American extreme vehicles are considered in detail by Leahy et al. (2014), and Chapter 6. There are significant differences, perhaps due to the US Federal Bridge Formula and the lower maximum legal weights that exist in the US. Leahy categorizes American extreme vehicles in three types, which he refers to as low loaders, mobile cranes, and cranes with dollies. Low loaders consist of a tractor and trailer and have one large inter-axle spacing, usually in the range, 8–13 m. Mobile cranes have a rigid body and closely spaced axles with relatively large axle loads. The mobile cranes in the US generally have fewer axles than in Europe and often rest the boom on a trailing dolly during travel, to allow the crane's weight to be spread over a greater length. Leahy also found low loaders and mobile cranes in European WIM data but also found crane ballast trucks and did not find mobile cranes with dollies. In Europe, crane ballast trucks often travel with cranes and carry the ballast needed to provide stability to cranes in service. They consist of a tractor and trailer units but do not have the single large spacing found in low loaders. Both mobile cranes and crane ballast trucks have a large load concentrated over closely spaced axles. As such, they are important for short-span bridge loading.

A comparison of vehicle single axle, tandem axle, and tridem axle weights in the US (FHWA 2018) and Europe (OBrien & Enright 2011) is shown in Figure 2.11. The histograms for the second axle load are significantly different. The tandem axle distributions are similar with a loaded vehicle weight peak of about 150 kN and unloaded of about 50 kN. For the European tridem axle data, a loaded tridem peak can be seen around 220 kN and unloaded around 60 kN. The same trend is absent in the US data which has only one peak around 130 kN.

In the development of live load models, usually the cumulative distribution functions are plotted on normal probability paper to emphasize the upper tails, which include the most extreme vehicles. Series of plots presenting GVW and axle load distribution are presented on probability papero to emphasize the variation in weight distribution and the importance of maximum load effects from a bridge design point of view.

GVW from several states in the US are shown in Figure 2.12. These plots are of unfiltered GVW data, so they include permit as well as standard vehicles. The heavy vehicles with GVW above 300 kN, which corresponds to about 30 tonnes, show significant variation. Data from three sites appear to reach an upper limit around 1000 kN (Alabama, Washington, and Rhode Island). This could be explained if the WIM systems did not record vehicles in excess of this weight, a feature of the software used for some WIM systems. Significantly, data from the Florida site does not appear to have reached an asymptote for the quantity of data considered, suggesting that even greater weights may be recorded if more data were analyzed. It should be noted that all WIM data is a function of the size of the database and it is the trend rather than just the particular extreme values that are important. It should be further noted that these upper tails can be strongly influenced by a small number of very heavy vehicles and the trends may not be the same if the survey is repeated. Clearly all of these extreme vehicles are abnormal, i.e. permit vehicles, and are not intended to be represented by a live load model for normal loading.

Tandem axle weight data are plotted on probability paper in Figure 2.13, where a tandem is defined as two axles spaced between 1.0 m and 2.5 m. There are many 5-axle (Class 9) vehicles in the US that consist of a steer axle and two tandems. It can be seen that the two tandem weight distributions are of similar shape. The maximum tandem weight observed is up to 500 kN.

A tridem is defined here as a combination of three consecutive axles with the distance between the first and third axles between 2.0 and 5.5 m. Tridem weight distributions are plotted on probability paper for various US

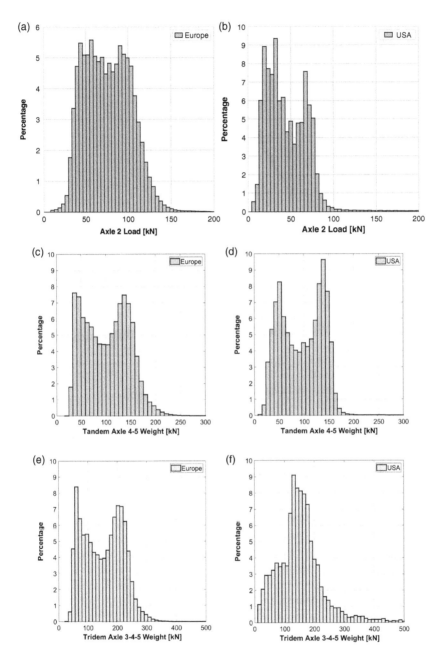

Figure 2.11 Histogram of axle 2 weight in a) Europe and b) USA; tandem axle 4-5 weight in c) Europe and d) USA; tridem axle 3-4-5weight in e) Europe and f) USA.

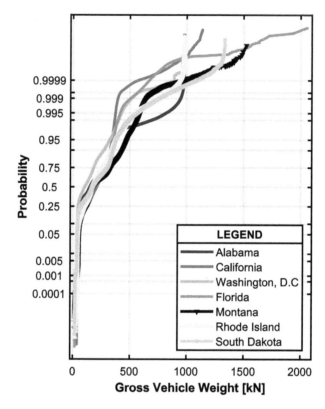

Figure 2.12 Normal probability paper plots of GVW data in different US states.

states in Figure 2.14. The heaviest tridem loads are observed in Alabama and Florida. The number of tridems identified in the California data is relatively low in comparison to other states, which indicates that such tridem configurations may not be typical in California traffic.

GVW data from five European countries: Slovakia, Poland, Slovenia, the Netherlands, and the Czech Republic, is plotted on probability paper in Figure 2.15. In Europe, legal weight and dimension limits vary between countries so the data can be expected to be more diverse than for the US. Confusingly, to ensure the free movement of freight, the European Union specifies a minimum value for the maximum legal GVW, i.e. member states cannot impose an unreasonably low upper limit that prohibits standard heavy trucks from passing on their roads. Except for the Netherlands' site, the distributions are consistent, and it is noted that vehicles in Europe are slightly heavier than in the United States (at least for the sites considered). The Netherlands has great volumes of very heavy trucks, which may be attributed to a dense population in a highly industrialized economy.

Tridem axle weight for Europe are shown in Figure 2.16. The tridem load distributions are similar for the sites/countries considered. It can be noted

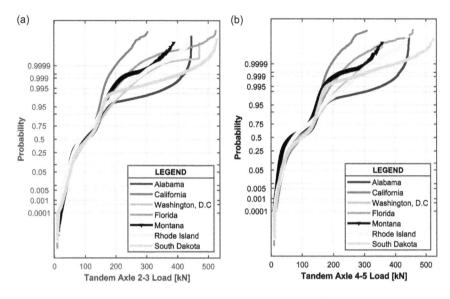

Figure 2.13 CDF of (a) tandem axle 2-3, (b) tandem axle 4-5 in the US.

that the tridem weights recorded in Europe are lighter than in the United States. The maximum tridem load is below 500 kN, whereas in the US, tridems up to 700 kN were recorded.

2.5 BRIDGE LOAD EFFECTS

The service life of a bridge depends on many factors such as traffic loads, natural hazards, quality of materials and labor, extreme events, etc. Traffic-induced loads may cause damage to a bridge by fatigue and may accelerate corrosion damage through the periodic opening and closing of microcracks. Every passage of a truck across a bridge creates one or more stress cycles in the structure. For steel bridges in particular, this can result in an accumulation of fatigue damage over time. Fatigue damage increases rapidly when vehicles are overloaded but, perhaps more significantly, overload increases the risk of failure. To maintain bridge safety, the load-carrying capacity must resist the load effects corresponding to the specified return period. Overload causes these load effects to increase which, in effect, increases the risk of failure.

A Load Effect (LE) is anything that is affected by load, but the most common LEs considered are bending moment and shear force. The LE due to an axle can be calculated using an influence line, defined as the response to a unit axle force at a point. Knowing the axle weights, the combined LE response to a vehicle is found by simply adding the effects due to each individual axle. As a truck passes over a bridge, it generates a bending moment

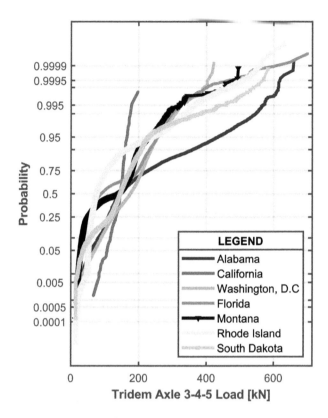

Figure 2.14 CDF's of tridem axle 3-4-5 weights in the US.

at each point along the span, and this moment changes as the truck crosses. At the critical point(s), the history of moment during the vehicle crossing is calculated and the maximum value identified.

Mid-span bending moment and support shear force effects are calculated in this way using the WIM data from the selected US states. An example of a 27 m (90 ft) long simply supported bridge is used to illustrate traffic load effects. These LE results are quite distinct from the raw GVW data considered in the previous section as the heavier vehicles tend to be longer so the moment will not increase in proportion to the GVW. The relationship between shear force and GVW is different again, as the shape of the influence line is different. Shear is more strongly influenced by single heavy axles or, in longer spans, by axle groups. Moment on the other hand, is more strongly influenced by GVW or, in shorter spans, by heavy axle groups.

Figure 2.17 shows the moment and shear effects plotted on normal probability paper. It can be seen that Rhode Island has some particularly large bending moments while South Dakota has some quite large shear forces. The largest bending moments and shear forces are caused by vehicles with heavily

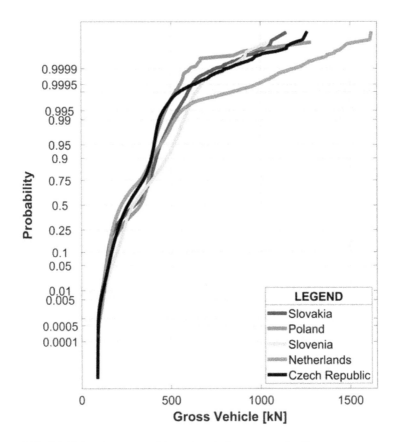

Figure 2.15 GVW distributions on Normal probability paper for European sites.

loaded and closely spaced axles. The same load effects are compared for the European data. In Figure 2.18 the CDF's for moment and shear effects are shown. The trends in the extreme moments are similar in all countries except for a smallnumber of outliers in Slovakia and Poland. For shear, the trends in allcountries are similar except for the Netherlands where they are greater.

2.6 FATIGUE DAMAGE

Traffic-induced loads may cause damage to a bridge by fatigue. Every passage of a truck across a bridge creates one or more stress cycles in the structural components, which results in the accumulation of fatigue damage over time. The passage of each heavy truck uses a certain amount of the fatigue life of the bridge. Bridges are subjected to variable amplitude stress cycles. The Palmgren-Miner (Miner 1954) rule provides a rational method to account for variable amplitude stress cycles. Miner's rule accounts for

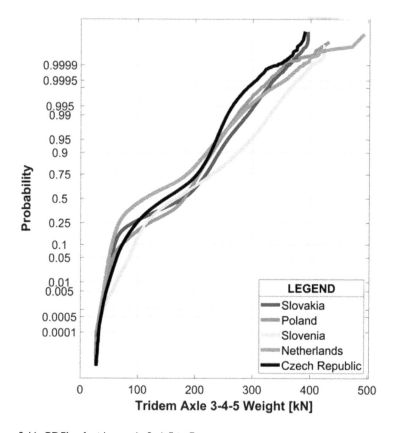

Figure 2.16 CDF's of tridem axle 3-4-5 in Europe.

the cumulative damage from a spectrum of applied stress ranges of variable amplitude. Using Miner's rule, an equivalent constant amplitude stress range, referred to as the effective stress range S_{eff}, and can be calculated by:

$$S_{eff} = \left[\Sigma \frac{n_i}{N} S_i^m \right]^{1/m}$$

(Eq. 2.1)

where:

n$_i$ – number of cycles at the ith stress range, S_i,
N – total number of cycles,
S_i – constant amplitude stress range,
m – fatigue exponent, structural and material dependent.

At a specific point along a bridge, the applied stress range can be determined by dividing the applied bending moment range by the section modulus. Hence, the available WIM data can be used to assess the fatigue damage due to the very large number of stress cycles experienced during the service life of a bridge.

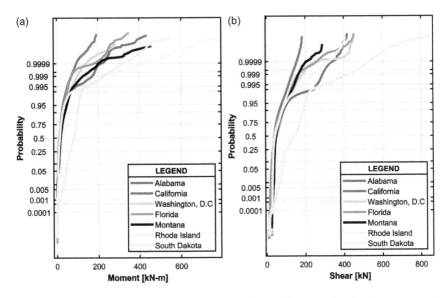

Figure 2.17 Normal probability paper plots of LE in a 27 m span bridge for a range of US states.

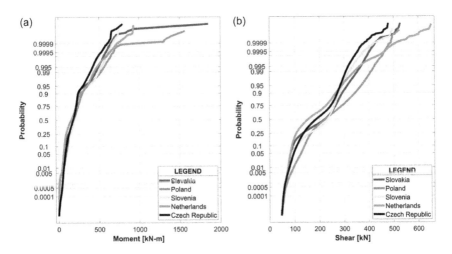

Figure 2.18 Normal probability paper plots of LE in a 27 m span bridge for a range of European WIM sites.

2.7 LEGAL LIMITS ON VEHICLE LOADS

Traffic consists of legal and illegal standard vehicles and permit vehicles. Legal limits for standard trucks are imposed to ensure the safety of transportation infrastructure and, for pavements in particular, to minimize damage due to heavy axle loads. In the US, federal law prevents states from imposing vehicle weight limits on interstate highways that deviate from established federal weight limits. This means that, for interstate highways, states are subject either to the standard federal weight limits or to state-specific 'grandfathered' limits or exceptions (US Code §127 1974), i.e. exceptions granted on the basis that they were in existence before the new law was enacted. Grandfather provisions, established in 1956, allow exceptions to the federal limits on axle weights and GVW and are particular to each state. The general situation worldwide is summarized in Figure 2.19. Standard vehicles are those that do not require a permit which, in some countries, include vehicles with grandfathered rights. In general, bridge design and assessment codes specify a notional load model for 'normal' traffic which is deemed to represent the extremes of standard vehicle loading. Vehicles seeking permits are compared to abnormal vehicles that the bridge has been found to have the capacity to carry.

Permit vehicles are those that require a permit because, according to the regulations on standard vehicles, they are oversized, overweight, or both. Permit vehicles need to follow the limitations specified in their permit, which may restrict the gross, single axle, and group axle weights. In the US, states have their own policies on the issuing of permits but must follow federal rules. Permits allow vehicles of specific configurations and sizes to exceed the standard vehicle size and weight limitations. Permits can be issued for single or multiple trips, usually referred to as special and routine permits, respectively. The permit may have limitations on designated routes, the number of trips, times of operation, and the necessity, or not, for escort vehicles. Illegally overloaded vehicles, with or without permits, belong to an unanalyzed portion of bridge traffic load that is more likely to create an extreme loading case.

Vehicle weights and dimensions vary greatly around the world. For example, Europe has very heavy crane ballast trucks which the US does not. On the other hand, the US has mobile cranes with dollies which Europe does not have. These differences between extreme European and US vehicles probably result from the US Federal Bridge Formula, which individual states are required to comply with. The primary purpose of the formula is to distribute vehicle load on highway bridges by limiting the axle

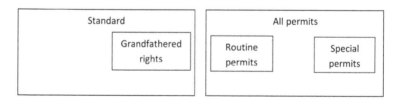

Figure 2.19 Vehicle categories.

configuration and axle load distribution. The formula limits the weight of any set of consecutive axles to:

$$W = 500 \left[\frac{3.28(LN)}{N-1} + 12N + 36 \right]$$

(Eq. 2.2)

where:

L – the distance between the outer axles of any group of two or more consecutive axles [m],

N – the number of axles in the set under consideration.

An exception is that two consecutive tandem axle groups are allowed to carry 15,423 kg each if the overall length of the four-axle set is at least 10 m (36 feet). Grandfathered rights vehicles are also allowed exceptions to the formula.

In the United States, the federal limit on GVW is approximately 40 tonnes. The GVW distribution for selected states is shown in Figure 2.20, in terms of the ratio of WIM vehicle GVW to the legal limit of 40 tonnes. Hence, all vehicles for which the ratio is above 1.0 can be taken to be permit or illegally overloaded. It can be seen that approximately 5% of vehicles exceed the legal limit. Florida, Montana, and South Dakota include very heavy vehicles that exceed the 40 tonnes threshold by three to four times. The Montana site is a notable exception, with 25% of vehicles above the legal GVW limit. It should be noted that this represents a large number of vehicles per day and increases the probability of multiple heavy trucks meeting or passing on a bridge.

In Europe, the legal limits vary widely between countries. For example, the GVW limit in Poland is 40 tonnes while in Sweden it is 60 tonnes. Figure 2.21 presents probability paper plots for GVW for the considered European countries, expressed as a multiple of a notional 50 tonnes limit. For most of the considered countries, approximately 1% of vehicles exceed 50 tonnes.

WIM data includes all vehicles and, given the relatively small number of permits issued, will include a small but important number of permit vehicles. In Figure 2.22 WIM data is compared to the permit vehicle database for Florida. The permit traffic is taken from the permit database that contains only vehicles that purchased and received permits in that state. As expected, the permit vehicles are significantly heavier. For example, at the 0.9999 probability level, the characteristic maximum GVW is around 700 kN for all vehicles, whereas for permit vehicles it is around 4600 kN. Of course, these weights are not comparable as there are clearly far more vehicles in the general population than in the permit vehicle database. It must also be noted that the extreme permit vehicles tend to distribute the load over many axles.

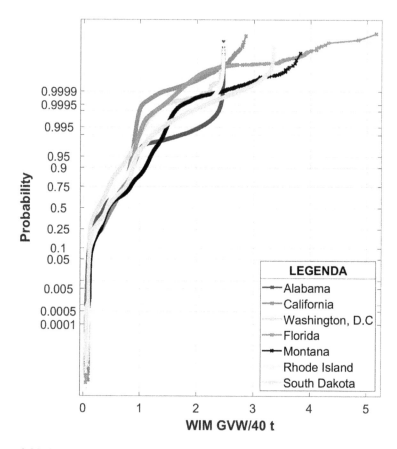

Figure 2.20 GVW ratios for various US sites on Normal probability paper.

Figures 2.23 and 2.24 show the tandem and tridem axle weight distributions from the WIM and permit databases (Ali et al. 2020). As expected, axle group weights are heavier in the permit vehicle database, but the difference from the axle group weights in the general vehicle population is much less pronounced than for gross weight. This supports the hypothesis that the US Federal Bridge Formula is requiring the gross weight of extreme permit vehicles to be spread over many axles whose individual weights are not excessive.

2.8 TRAFFIC LOAD FACTORS

A rational design or assessment of bridges requires a prediction of the expected maximum traffic load in the specified return period, i.e. corresponding to the specified level of safety. In AASHTO, the specified return period corresponds

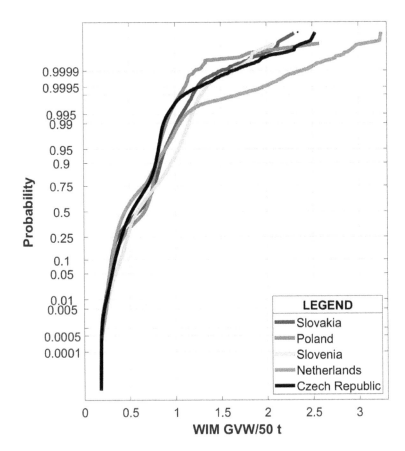

Figure 2.21 CDF's for on Normal probability paper of GVW ratio for various European
countries.

to the design life of the bridge, which is taken as 75 years (AASHTO 2014).
Thus, in AASHTO, the bridge is designed for the level of traffic load that
would be expected to be exceeded just once in its lifetime. In contrast, the
Eurocode suggests a design working life of 100 years and a return period of
about 1000 years so the bridge is designed for the level of load that would be
expected in just 10% of bridges in their lifetimes.

For most codes, load and resistance factors are calibrated so that the
structure can perform its function for its design life with a probability of
failure below the maximum acceptable level. In the current generation of
most design codes, the acceptability criterion for the minimum safety margin is specified in terms of a reliability index (Nowak & Collins 2013)
which is an indicator of the probability of failure. Load and resistance
parameters are treated as random variables and the reliability index is calculated accordingly.

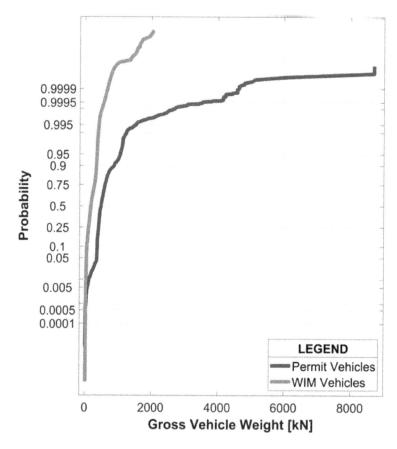

Figure 2.22 Normal probability paper plots of GVW for WIM and permit vehicle data in Florida.

The main source of information in deriving the statistical parameters of traffic load are the WIM measurements. Although substantial traffic data-bases have been assembled in recent decades, these are still insufficient to determine the maximum traffic load that a bridge may experience during the specified return period. Therefore, there is a need for an efficient and reliable technique to predict the extreme traffic load effect. Load and load-carrying capacity are both subject to considerable variation. Load-carrying capacity, or resistance, depends on material properties and the quality of workmanship, and there are uncertainties involved in the analytical model of the structure. This is why load and resistance are treated as random variables and partial safety factors applied to reflect their uncertainty. The derivation of these factors is referred to as calibration of the code. In a reliability-based calibration, the partial factors are chosen that satisfy predetermined reliability criteria, i.e. that provide a reliability index close to the target value.

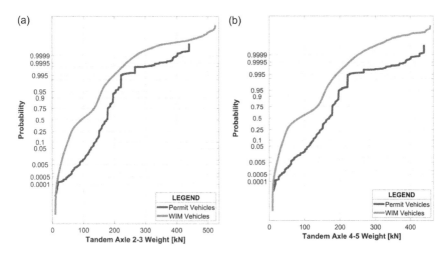

Figure 2.23 Normal probability paper plots of tandem axle weights for WIM and permit vehicle data in Florida, (a) tandem axle 2-3, (b) tandem axle 4-5.

The most important statistical parameters are bias factor (ratio of mean-to-nominal), coefficient of variation (ratio of standard deviation to the mean), and the type of the cumulative distribution function. However, highway traffic is strongly site-specific, not only from country to country but also within the country and even the local community. In addition, bridge safety is influenced by resistance as well as load so comparison of national design specifications requires a knowledge of not only design traffic load and its factors but also of resistance and its factors.

The first application of a reliability-based calibration procedure for bridge design was the derivation of load and resistant factors for the Ontario Highway Bridge Design Code in Canada (Nowak & Lind 1979). It was later applied to the calibration of the AASHTO LRFD Bridge Design Specifications (Nowak 1999). The basis for the current AASHTO LRFD Code (AASHTO 2014) was developed in the 1980s (Agarwal & Wolkowicz 1976, Nowak 1999). At that time, there was no reliable truck weight data available for the United States. Since then, an extensive database of WIM data has been collected by US states in multiple locations. The frequencies and weights of vehicles can change over time and this can affect the optimum load and resistance factors.

2.9 LIMIT STATES AND RELIABILITY INDEX

The current generation of design codes is based on a consideration of limit states. AASHTO defines four types of limit state: ultimate limit state (ULS), serviceability limit state (SLS), fatigue limit state, and extreme event limit

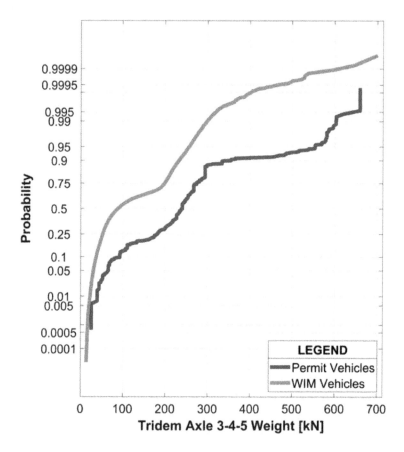

Figure 2.24 Normal probability paper plots of tridem axle weights for WIM and permit vehicle data in Florida.

state. ULS includes moment carrying capacity, shear capacity, axial compression, and axial tension. SLS includes cracking, deflection, and excessive vibration. The fatigue limit state is defined as when the number of load cycles reaches a limiting value. The extreme event limit state applies to earthquakes and other natural disasters.

Partial safety factors are applied to characteristic/nominal load effects to obtain factored design values. In many design codes, the SLS factors are taken equal to 1.00. For the fatigue limit state, it is important to know not only the magnitude of the traffic load but also the frequency of occurrence (numbers of cycles).

A mathematical representation of the border between acceptable (safe) performance and unacceptable (failure) performance is the limit state function. If R is a variable representing resistance or load-carrying capacity and Q is a random variable representing the load effect, then the limit state function can be given by:

$$g(R,Q) = R - Q \qquad \text{(Eq. 2.3)}$$

If the probability density functions (PDFs) for R and Q are as shown in Figure 2.25, then the PDF of $g(R, Q)$ is as shown and the probability of failure is represented by the shaded area. The state of the structure is then determined as safe if $g(R, Q)$ is zero or positive and unsafe if it is negative. It follows that the probability of failure, P_f, is the probability of $g(R, Q) < 0$.

As g is the difference of two random variables, R and Q, its mean is the difference of their means:

$$\mu_g = \mu_R - \mu_Q \qquad \text{(Eq. 2.4)}$$

where: μ_R and μ_Q are the mean values of resistance and load respectively. The standard deviation of the limit state function is:

$$\sigma_g = \sqrt{\sigma_R^2 + \sigma_Q^2} \qquad \text{(Eq. 2.5)}$$

where σ_R and σ_Q are the respective standard deviations. The reliability index, β, is defined as (Cornell 1968),

$$\beta = \frac{\mu_g}{\sigma_g} \qquad \text{(Eq. 2.6)}$$

Hence,

$$\beta = \frac{\mu_R - \mu_Q}{\sqrt{\sigma_R^2 + \sigma_Q^2}} \qquad \text{(Eq. 2.7)}$$

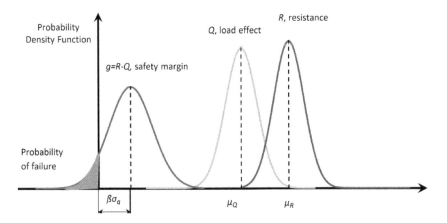

Figure 2.25 Probability density functions for load, resistance, and limit state function, g.

The probability of failure, P_f, can be calculated using the CDF of g. If R and Q are both normal random variables, then (Cornell 1968):

$$P_f = \Phi(-\beta) \qquad\qquad \text{(Eq. 2.8)}$$

where Φ is the CDF of the standard normal distribution.

2.10 EXTRAPOLATION OF IMPOSED TRAFFIC LOAD EFFECTS

Development of the design load requires a prediction of the maximum expected load effect. Total LE is a combination of load components. The basic load combination for bridges is imposed traffic load, L, and dead load, D:

$$Q = D + L \qquad\qquad \text{(Eq. 2.9)}$$

The statistical parameters for dead load are available in the literature. For in situ concrete, for example, the bias factor is in the range 1.03–1.05 and coefficient of variation is in the range 0.08–0.10.

Finding the statistical parameters for traffic load is more complex. Available WIM data covers much shorter time periods than the specified return periods of, for example, 75 or 1000 years. Therefore, the cumulative distribution functions obtained using the available WIM data have to be extrapolated in some way. In an early study, Nowak & Lind (1979) calculated 50-year characteristic maximum bending moment using data from a truck survey conducted by the Ontario Ministry of Transportation (Agarwal & Wolkowicz 1976). The CDF was plotted on semi-log paper (straight line would imply exponential statistical distribution) to emphasize the trend in the low-probability upper tail region.

In the development of the Ontario Highway Bridge Design Code, the vehicles from the survey data mentioned earlier (Agarwal & Wolkowicz 1976) were run over influence lines. Mid-span bending moments and shears were calculated from the truck weight data. The simply supported span lengths were varied from 3 to 60 m (10 to 200 ft). Then each LE was divided by the corresponding notional truck loading in the Ontario Highway Bridge Design Code (Nowak & Grouni 1994). The resulting moments and shears were plotted on normal probability paper. Then, the upper tail of the CDF was extended with an extrapolation line reflecting the trend. The extrapolation turned out to be close to a straight line (Nowak & Grouni 1994). It should be noted that, in tail fits of this kind, the last few points often deviate randomly from the trend. This is to be expected and is not significant as they represent a very small proportion of the whole data set.

The objective of extrapolation is to predict the expected maximum LE for an extended period of time. In the original calibration of the Ontario Highway Bridge Design Code, it was assumed that the economic lifetime of a bridge was 50 years and in calibration of the AASHTO LRFD Code, a lifetime of 75 years was considered. In both of these codes, the return periods were chosen to be equal to the design lives. The maximum value of LE corresponds to $1 - 1/(N+1)$, where N is the total number of vehicles (records) in the WIM data. If the time between vehicle crossings is t, then for any longer period of time, T, the number of vehicles is $N = T/t$. For example, if N vehicles are recorded in 1 year, $t = 1/N$, and, for $T = 50$ years, the number of expected vehicles is NT, which corresponds to a probability of non-exceedance of $1 - 1/(NT + 1)$ on the vertical scale. The expected maximum LE can be predicted by extrapolating the CDF from $1 - 1/(N+1)$ on vertical scale to $1 - 1/(NT + 1)$ and then reading the corresponding value of LE on the horizontal scale.

The procedure can be illustrated using WIM data in Figure 2.26. The number of measured vehicles (number of records) is $N = 365,000$ and the

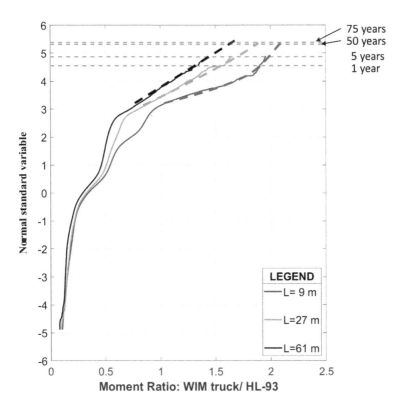

Figure 2.26 Extrapolation of the upper tail of the CDF of the moment ratio on Normal probability paper.

records were collected for 12 months. The measured vehicles were run over influence lines to determine the maximum bending moment. This was done for simple spans of 9, 27, and 61 m. Non-dimensional moment ratios were obtained by dividing by the corresponding HL-93 moments (AASHTO). The resulting moment ratios are plotted on normal probability paper. For each considered span length, the maximum value corresponds to $1 - 1/(N + 1)$, where $N = 365,000$.

The next step involves subjective judgement as the CDF trend needs to be extrapolated. The upper tail of the CDF is extended by what appears to be the best fit. The extrapolated tail of CDF serves as a basis for determining the expected maximum moment for any specified return period, T. For T = 50 years, the probability of exceedance is $1/(NT + 1) = 5.48 \cdot 10^{-8}$ and the non-exceedance probability on the vertical scale is unity minus this value. Similarly, for T = 75 years, the probability of exceedance is $3.65 \cdot 10^{-8}$, as shown in Figure 2.26. It turns out that the extrapolated upper tails are close to straight lines. The characteristic maximum 75-year moment is then determined as the moment corresponding to the non-exceedance probability of $1 - 1/(NT+1)$.

Assuming that the occurrence of heavy vehicles in consecutive years is a series of independent events, the CDF of the maximum moment for multi-year time periods can be derived from the CDF for a shorter time period. For example, if $F_1(x)$ is CDF for the maximum one-year moment, then the CDF for the maximum 75-year moment, $F_{75}(x)$ is (Nowak & Collins 2013):

$$F_{75}(x) = \left[F_1(x) \right]^{75} \qquad \text{(Eq. 2.10)}$$

Ghosn and Moses applied a multi-dimensional stochastic approach to develop a traffic load model based on WIM data collected in Ohio (Ghosn & Moses 1985). This approach takes into consideration the critical factors that directly affect the expected bridge traffic load such as multi-lane distribution, multiple presence, girder distribution, and future traffic growth factors. The maximum traffic load effect for a 50-year return period is determined as:

$$LE_{50} = a\,m\,H\,W_{0.95} \qquad \text{(Eq. 2.11)}$$

where:
  a – factor that depends on truck configuration and span length,
  H – headway factor,
  W_{95} – 95% characteristic value for dominating truck weight,
  m – variable that reflects the randomness in the axle configuration of representative random traffic.

They estimated the maximum LE (bending moment) acting on a single girder in a 50-year time period, as:

$$LE_{50_G} = LE_{50} \, g \, G_r \qquad\qquad \text{(Eq. 2.12)}$$

where:

g – girder distribution factor,
G_r – future traffic load growth factor.

This live load model was applied for assessment of the expected maximum traffic load for evaluation of existing bridges (Ghosn et al. 2010) and for the derivation of state-specific traffic load factors (AASHTO 2011).

The presented methods for prediction of expected maximum LEs are based on extrapolation and are defined without any reference to a specific vehicle(s): GVW, axle configuration, or axle loads. Such a vehicle(s) may be unrealistic or physically impossible. An alternative approach, based on traffic simulation using Monte Carlo techniques, has been applied by other researchers (Enright & OBrien 2013, Bailey & Bez 1999, OBrien et al. 2006, O'Connor & OBrien 2005) and can provide insights into the nature of extreme loading scenarios. Enright & OBrien in particular carried out simulations of thousands of years of traffic to identify the types of loading events that may govern at the level of the specified return periods.

2.11 TRAFFIC LOAD FACTORS

The design formula in the AASHTO code is typical of what is used in most modern codes:

$$\gamma_Q \cdot Q_n \le \phi \cdot R_n \qquad\qquad \text{(Eq. 2.13)}$$

where:

γ_Q – load factor,
ϕ – resistance factor,
Q_n – nominal load effect,
R_n – nominal resistance (load-carrying capacity).

The safety margins beyond the characteristic (nominal) level are provided by the load and resistance factors. The role of load factor is to increase the nominal load while the resistance factors decrease the design load-carrying capacity. Load and resistance factors are determined in the reliability-based calibration procedure. The optimum values of factored load and factored resistance can be found as the coordinates of the so-called design point (Nowak & Collins 2013).

If the limit state function is given by Eq. 2.3, and R and Q are independent Normal random variables, then the reliability index is given by Eq. 2.7. Then the coordinate of the design point for load can be determined from:

$$Q^* = \mu_Q + \frac{\beta\sigma_Q^2}{\sqrt{\sigma_R^2 + \sigma_Q^2}} \qquad \text{(Eq. 2.14)}$$

And the coordinates of the design point for resistance can be determined from:

$$R^* = \mu_R - \frac{\beta\sigma_R^2}{\sqrt{\sigma_R^2 + \sigma_Q^2}} \qquad \text{(Eq. 2.15)}$$

where:

Q^* – coordinate of the design point for Q,

R^* – coordinate of the design point,

But the factored load is equal to the coordinate of the design point for load, Q^*

$$Q^* = \gamma_Q \cdot Q_n \qquad \text{(Eq. 2.16)}$$

So, the load factor is

$$\gamma_Q = \frac{Q^*}{Q_n} \qquad \text{(Eq. 2.17)}$$

Load factors for dead load and traffic load can be calculated using the following equations by substituting statistical parameters for load and resistance components.

The coordinates of the design point for resistance can be determined from:

$$R^* = \mu_R - \frac{\beta\sigma_R^2}{\sqrt{\sigma_R^2 + \sigma_Q^2}} \qquad \text{(Eq. 2.18)}$$

and the resistance factor can be calculated from:

$$\phi = \frac{R^*}{R_n} \qquad \text{(Eq. 2.19)}$$

It is important to note that the calculation of load factors requires not only statistical parameters of load but also of resistance. Load factor selection cannot be separated from a consideration of resistance. If load and/or resistance are not Normal random variables, then the above listed equations can still be used but the results are approximate.

Chapter 3

Short-to-medium span bridges

Colin Caprani and Roman Lenner

3.1 INTRODUCTION

3.1.1 The physical and statistical phenomenon

Actual highway traffic loading on short to medium span bridges is characterized by the presence of uncertain single or multiple heavy vehicles on the bridge deck. The vertical static loading can be described in terms of the individual axle loads of each vehicle contributing to the total load effect, while interaction of the vibrating vehicle and bridge deck (see Chapter 4) leads to dynamic amplification of the static effect. This dynamic increase of the static load leads to higher load effects under free-flowing traffic. Congested, or stationary traffic, is slow moving, and so has little dynamic amplification. Consequently, it is a more critical loading situation for longer span lengths when multiple vehicles can be present (Caprani 2012). Within one lane, inter-vehicle gaps are crucial in determining whether single or multiple vehicles govern the extreme structural load effects (LEs) (e.g. bending moment, shear force). For multiple lane bridges, the density and composition of traffic controls the probability of side-by-side truck occurrences. This in turn governs the number of vehicles comprising a loading event, and hence the extreme LEs.

Identification of the extreme traffic-loading events is made more complicated when the structural form of the bridge is taken into account. Each bridge deck component has its own influence surface and these may be highly peaked or quite flat, either longitudinally and/or transversely. The transverse bridge deck stiffness for the component of interest therefore controls the significance of vehicle loading across adjacent lanes. Similarly, the longitudinal shape of the influence lines for shear, bending, or other LEs of interest is governed by the span count and connectivity type, whether a single simple span, two continuous spans, or multiple simple spans, for example.

Coupled with the vehicular arrangements and the structural system, engineering interest lies in some notion of an extreme loading situation, which reflects the underlying randomness of the phenomenon. Of interest

DOI: 10.1201/9780429318849-3

though, is not a topological arrangement of some prescribed and certain real vehicles, but rather the value of the traffic action (or LE) that has the prescribed probability of non-exceedance. Usually this is expressed in terms of some specified probability of non-exceedance or translated into a return period. It is important to recognize that the extreme loading events that can be observed daily, or even monthly, may not be a good guide to identifying those rare forms of loading events that can govern the extreme LEs at the return period of interest.

The interplay between traffic characteristics, traffic lanes, truck fleet, bridge length and structural form, level of dynamic amplification, and statistical extrapolation render the identification of the governing traffic loading for short- to medium-span bridges a challenging problem. There have been several paradigms in the approaches to this problem over the last century or so, but globally a data-driven approach has gained most traction in the last few decades. This chapter focuses on this approach, after some consideration of its historical development and some alternative approaches.

3.1.2 Load modeling approaches

Historically, until the 1960s or so, load models were developed empirically to reflect the total weight of a typical vehicle using the bridge, mostly corresponding to a local standard load including a safety margin. In the simplest terms, it was the LEs due to the weight of a typical carriage multiplied by a factor deemed to give the required safety. As transport distances increased due to improved vehicle technology, the empirical method was expanded to national or regional levels by imposing limits on individual vehicle weights and/or axle loads in order to have a functional stock of bridges. As traffic volumes began to increase, and the vehicle stock diversified, there was an evident shift in the associated loading on a national level (Dawe 2003). As a result, and simplistically stated, there was a demand for bridge design standards to introduce typical design loads for bridges.

Instead of standard formulas based on empirical estimates of weights and accompanying engineering judgment, methods based on actual traffic measurements started to emerge in the 1970s in Canada. This work was intended to reflect the frequency and variety of heavy trucks present in the traffic flow (Ghosn & Moses 1986, Harman & Davenport 1976, Nowak & Hong 1991). The inherent conservatism tied to the 'traditional methods' specified by codes (e.g. AASHTO 1977 which applied in Canada at the time) was deemed as obscure, and a rationale for a limit state based design code with live loads based on traffic data was reasoned as necessary (Csagoly & Dorton 1978). The Ontario Truck Survey of 1975 identified and weighed 9250 heavy trucks and formed the basis for the Ontario Highway Bridge Design Code load model (Agarwal & Wolkowitz 1976, Nowak 1994). In Europe, it was not until the 1980s that a concerted effort was made to measure truck traffic with the intention of deriving a bridge traffic load model (Calgaro & Sedlacek 1992). The Eurocode traffic load model was initially based on a set of pavement WIM data from Auxerre, France, in 1986. After consideration of similar data from other countries, this site further proved to provide a high percentage of heavy vehicles in comparison to the others, although a shortage of individual axle load distributions was noted (Hanswille & Sedlacek 2007). The final load model was derived based on the statistical prediction of extremes corresponding to a 5% probability of exceedance in 50 years. This WIM-based load modeling paradigm of solving the traffic-loading problem is now widespread and is considered in detail throughout this chapter.

An entirely different approach to traffic load model design was developed in Australia, where future load proofing of newly designed bridges was the main focus of the work. For this work, the maximum truck payload volume was determined from a consideration of road geometry for turning circles, clearance heights, and lane widths. The resulting available volume is then the maximum possible available for the transport of goods. Consideration then turned to the relative frequency and density distributions of bulk and volumetric freight transports. This, coupled with economic analysis, led to a target freight density of 0.73 tonnes/m³ from which the current SM1600 load model is derived. This provides for the design of new bridges to withstand future possible increases of gross vehicle weights. This philosophy embraces the fact that it is more cost effective to provide for higher loads at the design stage instead of strengthening existing bridges in future to withstand increasing legal load intensities (Heywood et al. 2000).

Overall, the initial efforts of deriving load models based on actual traffic data all suffered from unavoidable limited information at the time. The lack of data resulting from what would now be deemed rudimentary surveys and short measurement periods was then compensated for by necessary assumptions. This was to overcome the issues resulting from the lack of a full description of loading; for example, regarding the gaps between vehicles, multiple presence of vehicles in individual and adjacent lanes, variability of gross vehicle weights from site to site, or the influence of truck survey length on the resulting load effects. Most of the early assumptions about traffic characteristics have now been either refined by newly available data or redefined entirely as a result. This chapter provides guidance on the development of traffic load models for short- to medium-span bridges using measured traffic weight data.

3.1.3 WIM-based load modeling

Generically, in the development of a load model for structural design, it is either necessary to employ pragmatism and engineering judgment, or a more scientific approach of utilizing relevant available data. Given a history of recorded actions and the corresponding LEs, the prediction of a design load intensity for structures under similar loading is readily achieved through probability theory and statistical modeling. More recent decades have extended this to consider the distribution of resistances as well, so that the overall level of structural safety can be determined and managed.

The modeling of structural behavior has improved since the advent of the finite element method, so that the calculated LEs under a given load are now taken to be very close to those of the real structure. For traffic loading on bridges, this means that the main uncertainty comes from the traffic itself, and this data is the main input to the scientific approach to developing a load model. The static response of the structure under load can be represented through influence lines and surfaces for the LE of interest. When coupled with the measured traffic, this forms the basis of modern WIM-based traffic load modeling. The detailed process is summarized in Figure 3.1, including the statistical modeling to determine the extreme LE and the subsequent design of a notional load model where required. The

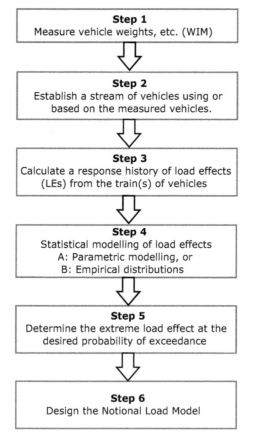

Figure 3.1 Flowchart of the main steps in WIM-based development of a notional load model.

input data is cleaned and calibrated in Step 1 to produce a suitable database of measured vehicles, containing axle weights, numbers of axles, and inter-axle spacings. A train of this data, including inter-vehicle spacings, is established in Step 2. There are some studies where a 'train' of vehicles can be simulated but if a sufficient quantity of WIM data is available, this is not necessary (Fu et al. 2010). There are also intermediate bootstrapping approaches that can be adopted (Melhem et al. 2020), that is, random sampling from existing data. The bridge response to the train of vehicles (either recorded or simulated) is calculated in Step 3 to give a history of LEs on the bridge. Generally, a suitable statistical distribution is fitted to this LE data in Step 4 and the characteristic maximum value is found in Step 5. However, if a sufficient quantity of artificial LE data can be generated, then fitting to a distribution may not be necessary and the value corresponding to the characteristic maximum may be obtained directly (Enright & OBrien 2013) – Step 4B. Lastly, for where it is required, a notional load model is developed in Step 6 to represent the characteristic loading due to traffic but through a simplified representation of the loading (e.g. uniformly distributed load and axle-bogey or design truck).

The following sections of this chapter examine each of the steps of Figure 3.1 in detail. Section 3.2 outlines the aspects of WIM that must be considered for the load modeling, in conjunction with Chapter 2. Step 2 is considered in Section 3.3, including consideration of multiple lanes. Detailed consideration of the calculation of the LEs under the traffic streams (Step 4) is given in Section 3.4. Section 3.5, in conjunction with Chapter 4, describe the means of allowing for dynamic interaction which, depending on the approach, contributes to Step 3 or Step 4. Section 3.6 examines the statistical prediction of the extreme loading and provides a means of identifying the governing form of traffic allowing for dynamic interaction. Finally, for Step 6, Section 3.7 outlines some approaches where it is required to develop a notional load model.

3.2 TRAFFIC DATA

3.2.1 WIM data and recordings

Chapter 2 provides a detailed examination of WIM technologies and the resulting data. For application to bridge traffic-loading estimation, the essential characteristics to measure in the vehicle stream are:

- Date of record;
- Time of vehicle arrival;
- Vehicle speed;
- Gross vehicle mass;
- Individual axle loads;
- Axle spacings.

Further useful information such as the bumper-to-axle gaps can be obtained at some sites, and the vehicular configuration, coded according to the pertinent vehicle classification system, can assist with identifying concatenation errors in the WIM data.

While the accuracy of WIM measurements can vary, O'Connor & OBrien (2005) point out that bridge loading estimation is not very sensitive to WIM accuracy for the development of load models up to 50 m. Nevertheless, the collected data should be properly cleaned and calibrated as described in Chapter 2 to ensure that an accurate representation of the actual traffic is used in the analysis.

3.2.2 WIM data filtering and cleaning

Once the calibrated WIM data is obtained, it is necessary to apply a basic filtering process to remove potentially incorrect values that could distort the results. These would typically include negative axle spacings, high weight differences in adjacent axles of a group or excessive speeds. Chapter 2 describes some of these approaches in detail. Calibration methods that can be applied to the data after collection are particularly valuable, such as the Truck Tractor method (TT method) developed by de Wet (2010) in South Africa for longer 6- and 7-axle vehicles. Such methods suppress systematic biases associated with each WIM station after the data collection. While a single calibration factor is generally applied to all axle load measurements, some methods have been developed where the systematic error is removed based on different factors, such as speed (Papagiannakis et al. 1995); for further techniques refer to Enright & OBrien (2011).

The time resolution of a WIM station is a critically important parameter for the development of load models as it governs the gaps between vehicles in any subsequent analysis. WIM system time resolutions can range from 0.001 s to 1 s depending on the equipment used. For bridge traffic loading it is recommended to have a time stamp resolution of 0.01 s or smaller (Enright & OBrien 2011). Furthermore, when multiple lanes are measured at the site, high resolution is required in order to obtain the correct spatial distribution of all trucks occupying all lanes.

3.2.3 Measurement duration and extent

The duration of available WIM measurement dictates the methods used in later steps of the bridge traffic load estimation analysis (Figure 3.1). Of course, longer measurement periods yield more information about the vehicles and variations of the traffic flow. Specifically, longer durations allow the characterization of truck traffic over weekends, or during school holidays. Furthermore, seasonal or other regional variations (e.g. harvest time) in the truck traffic can significantly influence the development of a load model. Depending on the method of statistical extrapolation adopted (Section 3.6), longer or shorter periods of WIM data may be required. Often, block maximum values are often analyzed, i.e. the maximum LE in a specified block of time. It is important that these blocks are similar in nature and so are independent and identically distributed (Basson & Lenner 2019). For example, for many sites, it is reasonable to assume that working days are similar but that weekend days are different. In general, it can be taken that long periods of measurements are desirable, made up of statistically similar time blocks. In the absence of long periods of measurements, advanced modeling techniques can be applied (Caprani 2013).

For the development of load models covering more than one site, such as regional, or national load models, variations in traffic across the region should be captured by a geographically distributed WIM station network. For example, the composition of freight and traffic will vary significantly between sites on key inter-urban routes, those proximate to industrial sites, or those near a commercial port. Further, temporary spikes in loading can occur at some locations due to construction activity. Some previous code development has selected the WIM station exhibiting the highest intensity of heavy vehicles in terms of both the flow and the loading as this should result in a conservative load model for the region in question (Sedlacek et al. 2008). However, aggregating multiple WIM stations would be more statistically appropriate for the entire region and could be considered through appropriate statistical mixing.

3.2.4 Overloaded and permit vehicles

The distinction and nuances involved in considering standard, illegally overloaded, and permit trucks is detailed in Chapter 2. Here, a typical example facing the analyst is illustrated using data recorded at Roosboom station in South Africa. Figure 3.2 shows the overall site histogram of GVW, along with that of each vehicle type, as identified through the number of axles. At this site, 56 t is the legal limit and there are clearly 7-axle vehicles operating

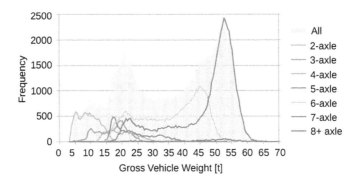

Figure 3.2 GVW frequency distribution by vehicle type at Roosboom (Lenner et al. 2017).

at weights beyond the legal load. The bi-modal shape of the GVW distribution is typical, and corresponds to the loaded and the unloaded vehicles. The second peak at approximately of 53 t is characteristic of freight operators staying below the legal limit while maximizing transportation.

It is virtually impossible to distinguish between illegally overloaded vehicles and permit vehicles from WIM data alone, such as that shown in Figure 3.2. Identifications at the WIM site can be improved through video surveillance, or ideally automatic number-plate recognition. For the development of 'normal' traffic load models, it is desirable to filter the permit vehicles, as these are more appropriate to an 'abnormal' traffic load model. Nevertheless, the illegally overloaded vehicles do form part of the population of normal traffic, and must be accounted for in its statistical description.

3.3 LOADING EVENTS

The data collected at WIM stations does not provide direct information about the loading a bridge might experience. Rather, it can be utilized in static calculations of LEs. For computational efficiency, this is typically done by passing individual axle loads over corresponding influence lines. These axle loads may be those directly measured, or they can be simulated based on the measurements using Monte Carlo or bootstrapping methods. Either way, the aim is to obtain bending moments and shears representative of those experienced by a bridge at that site.

3.3.1 Direct use of measured WIM data

When there is a sufficiently long record of WIM data available, it can be used directly to obtain bridge LEs. The main benefit of this approach is that any statistical relationships between the variables comprising the truck fleet and traffic are implicitly captured in the resulting LEs. For example, there is correlation between the vehicle weight, number of axles, axle weights, and vehicle lengths, which can be difficult to otherwise model, as explained later. A drawback of using WIM data directly, is that alternative situations cannot be examined; for example, hypothesized traffic growth or modal shifts (see Chapter 6) cannot be studied.

A vehicle in free-flowing traffic travels at around 25 m/s and therefore crosses a short- to medium-length bridge in one or two seconds. As gaps between vehicles are generally greater than one second, the number of occurrences of multiple following vehicles on a short- to medium-length bridge tends to be small. Even where it does occur, the second vehicle is often only partially on the bridge when the first is at a critical location and is sometimes presumed to contribute little to the LE. However, the contribution depends, of course, on the shape of the influence line or surface.

In any case, for single lanes, some researchers consider only single vehicle loading events in the calculation of characteristic loads. This approach is especially suitable for shorter bridges. However, it is not difficult to consider the possibility of multiple in-lane vehicle presences when processing WIM data, as weights, speeds, and relative positions (arrival times) are known for all vehicles and can be taken directly from the data. Consequently, a long convoy of vehicles can be built from the underlying WIM data. The same is true for two-lane bidirectional bridges, where the train of vehicles in each direction are considered (usually independently). The case of multi-lane same-direction traffic is more complex and is discussed in Section 3.4.

When considering multiple in-lane vehicles, it is important to prevent unrealistic overlapping of vehicles during the bridge crossing. While this can be avoided by modeling driver behavior, including acceleration and deceleration (as is discussed for long-span bridges in Chapter 5), the computational effort involved is unwarranted for the short durations involved. Nevertheless, any overlapping must be prevented. One approach is to take the times of arrival as being those at the WIM system, and to impose a constant velocity on all vehicles during the bridge-crossing event. In most cases, where speeds do not vary much, taking the average velocity of the stream, and imposing it on all vehicles, is adequate. However, in non-stationary conditions, such as transitions from free-flow to congested flow, which may exist in the measurements, speeds can fluctuate significantly. In these cases, if the selected constant velocity is higher than the actual velocity of several adjacent vehicles, then the space headways will be overestimated, leading to underestimated LEs. In contrast, for brief periods of velocities higher than the set constant velocity, the problem is reversed.

Another means of preventing overlapping is to consider each successive arrival in turn, adjusting the velocity of the following vehicle as necessary. Key dimensions of the vehicles for this situation are illustrated in Figure3.3, and the times of arrival are based on the successive arrivals of the front axles (thereby including the length of the lead vehicle). In particular, the velocity of the following vehicle is adjusted to ensure that a target time gap to the vehicle in front is ensured at the end of the crossing. In the following, subscripts f and r refer to the front and rear vehicles, and L and

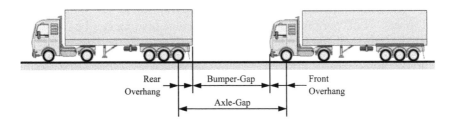

Figure 3.3 Key physical parameters of the spacing of vehicles.

v to their lengths and velocities respectably. Of course, when $v_r \leq v_f$ there is no risk of overlapping on the bridge, and so what follows applies when the rear vehicle is catching up, $v_r > v_f$. At the end of the crossing, the target minimum time gap (headway) to the vehicle in front can be written as:

$$\Delta T_{min} = \frac{L_f + b_S}{v_r} + b_T \qquad \text{(Eq. 3.1)}$$

where b_S is a physical space buffer (e.g. wheel diameter, say 1.0 m) and b_T is a time buffer (e.g. extreme minimum feasible time gap between vehicles, e.g. 0.1 s conservatively). These buffers can be further adjusted, depending on whether or not the WIM data contains the front and rear overhangs (Figure 3.3). If the headway between the vehicle arrivals at the start of the bridge of length L_B is Δt (i.e. the time gaps in the WIM record), then this minimum time gap at the end of the traverse will not be achieved when:

$$\Delta t + \frac{L_B}{v_r} - \frac{L_B}{v_f} < \Delta T_{min} \qquad v_f \geq v_r \qquad \text{(Eq. 3.2)}$$

When this is the case, to achieve the target minimum time gap, the speed of the rear vehicle must therefore be adjusted to:

$$v_r^* = \frac{L_B}{L_B/v_f + \Delta T_{min} - \Delta t} \geq v_f \qquad \text{(Eq. 3.3)}$$

but is bounded to the speed of the front vehicle. In considering relationships like this, it is beneficial to draw the space-time plot of vehicle movements (straight lines for constant velocity) on the bridge deck.

3.3.2 Generation of artificial traffic streams

Where it is judged that there is insufficient WIM data to obtain sufficiently long periods of LEs, artificial streams of vehicles can be generated numerically. A benefit of this approach is that different scenarios of traffic flow and composition can be studied, including potentially traffic growth. However, the main challenge is the accurate modeling of the complex relationships that exist within the vehicle and traffic flow parameters.

The important parameters of traffic for bridge loading can be classified as (1) the physical parameters of the vehicles, and (2) the parameters of the traffic flow. With this classification, it is useful at times to think of a 'garage' of vehicles that are inserted into defined traffic streams. The physical parameters of a vehicle are the number of axles, the axle spacings and weights, and the gross vehicle weight. The parameters of the traffic flow include the hourly flow in each lane, the distribution of vehicle types between lanes, and the headway or gaps between successive vehicles. For short- to medium-span bridges, it is reasonable to neglect the possibility of lane changes on the bridge, and so the measured traffic properties in each lane at the site of interest is sufficient. Consequently, statistical independence of traffic between lanes is commonly assumed in the literature and is a reasonable assumption for traffic in opposing directions. However, it is known to be inaccurate for same-direction traffic (OBrien & Enright 2011). Where there are multiple same-direction lanes, light trucks often overtake heavier ones, resulting in correlated weights. Truck presence is also correlated – the presence of a truck in the overtaking lane is almost always associated with a vehicle in the slower lane. These issues are addressed in Section 3.3. The composition of the traffic stream is usually determined from the measured proportions of the different vehicle classes, however defined. Monte Carlo simulation is used to generate the class of the 'next' vehicle in the stream using an appropriate model. Typically, independence of vehicle class arrivals is assumed, but autocorrelations of vehicle class within the stream can be modeled where necessary.

Considering the 'garage' of measured vehicles, each vehicle classification has its own statistical model, used to generate vehicles representative of that type. The classifications can be quite simple, such as the number of axles, or more complex (e.g. formal FHWA, Euro13, or AustRoads Classifications). A problem with this, is that only classifications with enough samples in the WIM data are modeled, thereby artificially limiting the vehicle types in the stream. Notably Enright (Enright et al. 2016, Enright 2010) overcomes this problem and provides a means of generating vehicles with more axles than those observed in the traffic stream. However, this method does require considerable traffic data (e.g. one year) to be reasonably robust. Vehicle articulation can also be used to distinguish vehicle classes for statistical modeling. For example, Bailey (1996) uses 14 classes of vehicles, but only considers units with up to five axles, the remainder being made up of tractor-trailer combinations.

3.3.2.1 Monte Carlo simulation

Harman & Davenport (1976) were among the first to propose Monte Carlo simulation of vehicles and headways to generate LEs for further statistical analysis. Monte Carlo simulation is a process that generates 'typical' data, consistent with a specified statistical distribution. Monte Carlo simulation can be based on fitted parametric (e.g. normal), or non-parametric (e.g. cumulative frequency) distributions. For example, if vehicle gross weight is known to be Normally distributed with mean of 50 t and standard deviation of 5 t, repeated application of this formula will generate numbers that are mostly between 40 and 60 t, with a concentration around the mean of 50 t. The benefit of parametric modeling is that samples beyond the WIM data can be generated, but are still reflective of the population. In contrast, when adopting non-parametric modeling (also known as bootstrapping), no sample beyond that already observed can be generated. Consequently, non-parametric modeling can be suitable for axle spacings and axle weights as a proportion of GVW, but at least for the extreme upper tails of GVW, parametric modeling is preferable as it allows sampling beyond the measured data.

3.3.2.2 Modeling vehicles

A basic set of statistical models of the relevant vehicle properties is summarized in Table 3.1. More advanced models for some of these properties are reported in the literature, and some of these are summarized below.

The fitting of bi-modal and tri-modal Normal distributions is a flexible tool for modeling a range of properties including gross-vehicle weight, axle spacings, and axle weight. For Gross Vehicle Weight, Harman & Davenport (1976) were the first to use a linear combination of three Normal distributions – the tri-modal Normal distribution, given by:

$$f_X(x) = \sum_{i=1}^{3} \rho_i \, \varphi(x; \mu_i, \sigma_i) \qquad \text{(Eq. 3.4)}$$

in which i is the mode, ρ_i is the weighting, and $\varphi(x; \mu_i, \sigma_i)$ is the probability density function for a Normal distribution with mean, μ_i and standard deviation, σ_i. This model has been widely adopted.

Bailey (1996) uses the Beta distribution for each of the traffic characteristics. This is a suitable distribution, as it is very flexible and has upper and

Table 3.1 Statistical Models of Vehicle and Flow Characteristics (Caprani 2005)

Traffic Characteristic	Statistical Model
Gross Vehicle Weight (GVW)	Tri- or bi-modal Normal distribution
Axle spacings	Uni- or bi-modal Normal distributions, as appropriate
Axle weights for 2- and 3-axle trucks	Tri- or bi-modal Normal distributions, as appropriate
Axle weights for 4- and 5-axle trucks	Expressed as a percentage of GVW for the first and second axles and for the remaining tandem group. In each case, the percentage is modeled as a Normal distribution
Composition	Measured percentage of 2-, 3-, 4-, and 5-axle trucks
Speed	Normal distribution – considered independent of truck type and uncorrelated with GVW
Flow rates	For each hour of the day, the average flow rate (ignoring weekend days and public holidays) can be used for all the days available
Headway	Modeled with a number of distributions dependent on flow and gap, as described in (OBrien & Caprani 2005)

lower limits which can be used to impose physically realistic values. Bailey considers axles within groups as having equal weight, since the weight is evenly distributed between closely-spaced axles. A generalized bi-modal Beta distribution is used to fit the observed axle group weights, and correlation of this weight with the GVW is also considered.

To better simulate individual axle loads, Crespo-Minguillón & Casas (1997) allocate axle weights and GVW based on their measured correlations. Geometries are based on measured correlation coefficients for axle spacings. The GVW and axle weight distributions are defined numerically from measured cumulative distribution functions derived from the histograms of WIM data (i.e. a non-parametric approach). Srinivas et al. (2006) use copulas to accurately model the dependence structure between multimodal distributions of axle weights. In a simplified manner, (Pérez, Sifre & Lenner 2019) use bivariate copulas to establish a relationship between the GVW and individual axle loads for up to 7-axle vehicles, that can be further utilized in the non-parametric approach of simulations.

Enright et al. (2016) use a semi-parametric approach for modeling some vehicle parameters. Up to a certain GVW threshold, where there is sufficient data, a measured bivariate histogram is used to generate GVW and the number of axles. The frequency threshold is selected as that level for which the bin count exceeds 16 observations, using a bin size of 1 t. Beyond this point, a parametric fit is used, in this case the upper tail of a Normal distribution. This semi-parametric approach means that average vehicle properties are exactly represented while ensuring that extreme weights can be generated beyond anything recorded in the database. For axle spacings, Enright et al. (2016) use empirical distributions of the ordered axle spacings and their positions in the vehicle; an approach which gives the axle layout for any vehicle and allows the generation of vehicle classes beyond the observations. For standard vehicles (e.g. 5-axle semi-trailer), this approach gives a similar accuracy to past approaches in which axle spacings are modeled based on position only. However, for heavier and less common vehicles (e.g. cranes, low-loaders, permit vehicles), Enright's approach is superior.

3.3.2.3 Generating gaps

Probably the most important parameter in the generation of artificial traffic steam for short- to medium-length bridges is the headway, Figure 3.3, as it controls the number of axle loads that can simultaneously exist on the bridge (OBrien et al. 2006). Headway is related to the physical gap by the vehicle length and overhangs and can be measured either in time or distance. WIM measurements typically contain headway in units of time and so many models represent it in this manner. The most basic vehicle arrival model is based on the assumption that vehicle arrivals are Poisson-distributed, resulting in a negative exponential distribution for headway.

Allowing for vehicle length, this becomes the shifted negative exponential distribution with cumulative distribution function:

$$F(\Delta t) = 1 - \exp\left[-\gamma(\Delta t - \Delta t_{\min})\right] \tag{Eq. 3.5}$$

where $\gamma = Q/(Q\,\Delta t_{\min} - 1)$ and Q is the flow (vehicles/hour). The minimum headway, $\Delta t_{\min}$ in this model is determined by a nominal vehicle length and speed. The criticism in Section 3.3.1 applies to the choice of these nominal parameters controlling vehicle gaps, and so an alternative model often used to avoid these problems is the Gamma distribution. However, this distribution gives a high probability to small headways. Interestingly, Crespo-Minguillón & Casas (1997) and Grave (2001) find that the distribution of 'normalized headway,' defined as the headway divided by the mean headway per hour, is consistent, regardless of flow. The cumulative distribution function for headway, Δt, is therefore given by:

$$F(\Delta t) = \frac{Q_T}{3600}\left[1 - e^{-\lambda \Delta t}\right] \tag{Eq. 3.6}$$

where λ is the mean normalized headway and Q_T is the truck flow (trucks/hour). Using this model for headways over 4 s, OBrien & Caprani (2005) propose a headway model which uses polynomial fits to the measured WIM data for the flow-dependent cumulative distribution function of headways less than 4 s (Figure 3.4). Enright (2010) also adopts this approach but instead of headway uses the axle-gap (Figure 3.3) as the parameter being modeled. This is useful because it avoids the potential for vehicle overlap on the bridge. Finally, the traffic flow rate is a particularly significant parameter as flow tends to vary through the day and clearly has a significant influence on gaps: (Eq. 3.6). To allow for this, it is essential that each hour of the day is simulated separately.

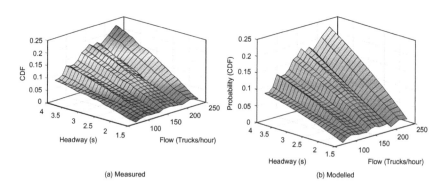

(a) Measured   (b) Modelled

Figure 3.4 Cumulative distribution functions for truck flows between 1.5 s and 4 s (OBrien & Caprani 2005).

When generating headways (and not axle-gaps) for successive arrivals, the minimum gap at the end of the crossing, shown in Equation (3.1), should be met to avoid inappropriate overlapping. Considering that the vehicle properties are already defined, including velocities, when the following truck is moving slower than the front truck ($v_r \leq v_f$), the minimum required headway upon arrival at the bridge is just this target gap $\Delta t_{min} = \Delta T_{min}$. However, when the following truck is catching up on the lead truck, using the notation from earlier, the minimum headway upon arrival at the bridge must be:

$$\Delta t_{min} = \frac{\Delta T_{min} v_r + L_B}{v_f} - \frac{L_B}{v_r} \geq b_T \qquad \text{(Eq. 3.7)}$$

Upon generation of the random headway according to the adopted headway model, the requirement of Equation (3.7) should be checked and imposed if necessary. For bridges with bi-directional traffic, the above formulation represents a 'mixed-datum' case in which the arrival datum is at one end of the bridge for the lanes in one direction, and at the other end for the other traffic direction. Caprani (2005) considers the minimum gap requirements for the single-datum case, where the vehicle arrival datum for both directions is located at one end of the bridge. While this is a more faithful representation of the single point of measurement of a WIM station, the additional complexity is rarely warranted for short- to medium- span bridges for which traffic parameters can be considered as a statistically stationary process over this short distance.

3.3.3 Traffic loading in multiple lanes

When considering more than one traffic lane, the analysis and modeling varies, depending on whether the lanes are assumed with traffic running in the same or opposing directions. Studies rarely consider more than two lanes, for two reasons: (i) for most short/medium-span bridges, the transverse stiffness is such that traffic in a lane rarely has a significant influence on lanes that are not immediately adjacent and (ii) trucks are often restricted from travelling in more than two lanes.

3.3.3.1 Opposing-direction traffic

Two lane bi-directional traffic is the most basic two-lane bridge to study. This is because the properties of vehicles traveling in opposite directions can be assumed to be statistically independent – there is no reason to expect the weights of two vehicles meeting on a bridge to be correlated. Hence, for this case, the generation of the traffic streams in each direction reduces to the generation of two independent lanes of traffic. While each direction may have its own vehicle characteristics, it is more common to consider just differences in flow and composition. For bi-directional bridges with more than one lane in one or both directions, the assumption of independence in direction usually remains valid, and then the problem reduces to consideration of two same-direction lanes.

3.3.3.2 Same-direction traffic

Two or more lanes of same-direction traffic are best examined considering just two lanes initially. In countries with asymmetric passing rules (e.g. Europe), experience suggests that the vehicles driving in the overtaking lane tend to be faster and therefore lighter than those in the driving or slow-lane. Consequently, in these situations, there are statistical relationships between lanes that should be captured for accurate representation of traffic loading. For countries with symmetric passing rules (e.g. the United States), it may be feasible to consider each lane as independent for some suitable sites. Chapter 5 discusses more about the influence of passing rules on bridge traffic loading.

OBrien & Enright (2011) review early work on same-direction multiple-lane bridge traffic loading. With little or no available data, early work was based on subjective assumptions. As an example, Nowak (1993) assumes that one in fifteen heavy trucks is part of an overtaking event. For these overtaking events, he goes on to assume that the truck weights are fully correlated in one of 30 cases. Kulicki et al. (2007) acknowledge the lack of sufficient data and that their assumptions are based on limited observations. For weight correlation, their assumptions are based entirely on judgment. Sivakumar et al. (2007) propose multiple vehicle presence probabilities that are a function of average daily truck traffic, as clearly heavier traffic results in more overtaking events. They use subjective field observations to find conservative values. Soriano et al. (2017) however use WIM data to address simultaneous loading and side-by-side effects, instead of the early assumptions. Multiple WIM sites from New York city are used to predict the effect of two combined lanes using a convolution based on the WIM records showing no correlation between trucks in the same or adjacent lanes. A similar approach is used by van der Spuy et al. (2019) where multiple lane records of both same- and opposite-direction traffic are used for the prediction of combined LEs in any number of lanes, given that multiple lane data were available. Interestingly, for the Eurocode load model (2003), each lane was simulated independently based on both direct use of the WIM data and Monte Carlo generation of artificial traffic streams (Bruls et al. 1996, Dawe 2003). Consequently, there was no explicit consideration of the relative positions of vehicles in adjacent lanes.

For load models that are based on the probability of side-by-side truck occurrences, as in the United States, a precise definition of 'side-by-side' is warranted. After a parametric study, Sivakumar et al. (2007) define a side-by-side event as one where adjacent trucks have a headway difference of less than 18.3 m (60 ft): a typical tractor-semi-trailer in the United States is 70 to 80 ft. This approach is subsequently refined with a description of side-by-side, staggered, following, and other multiple-presence events (Sivakumar et al. 2007). They use WIM data to estimate the frequency of these event types for different truck traffic volumes and bridge lengths. They also calculate multiple-presence probabilities either directly from WIM data or estimate them from traffic volumes.

3.3.3.3 Scenario modeling

OBrien & Enright (2011) propose the concept of 'scenario modeling' that addresses most of the challenges posed by multiple lanes of same-direction traffic. Based on the recorded data, they show that there are correlations between vehicle weights, speeds, and inter-vehicle gaps. In a study of data from two European sites it is found that, for nearly 75% of fast lane trucks, there is an associated truck within 2.0 seconds of it in the slow lane; i.e. most fast lane trucks are overtaking a slow lane truck. This phenomenon increases the probability of multiple-truck presence on a bridge over any simulation that treats the lanes as independent. Due to the difficulty of simultaneously modeling all the interdependencies in a traffic scenario, they propose to use the measured scenario as the basis for simulation.

An example of a traffic scenario taken directly from measured WIM data is shown in Figure 3.5. This traffic scenario has 21 parameters – the GVWs and speeds of seven trucks, six gap values, and a flow rate. The correlations and relationships between these parameters are implicit within each scenario, as they are taken directly from the measured WIM data. To allow for the possibility of loading scenarios other than those in the database, OBrien & Enright (2011) 'perturb' the scenarios identified from the WIM data before finalizing them, i.e. they make small changes to the key parameters by adding statistical 'noise.' They argue that these random perturbations do not significantly damage the integrity of the relationships between the parameters, yet allow for unforeseen and unrecorded combinations of vehicles. Notably, in a subsequent study, OBrien et al. (2015) validate the process by comparing it with a reference dataset generated by micro-simulation (Chapter 5).

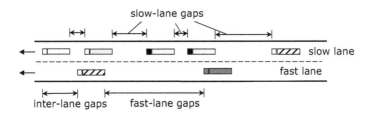

Figure 3.5 Traffic scenario (after OBrien & Enright 2011).

3.4 LOAD EFFECTS

Given a stream of vehicles, the induced structural effects due to this loading are of course the main goal of the computation. There are a wide range of methods for calculating these LEs, ranging from influence lines, to influence surfaces, to grillage models, to elaborate finite element models. Due to the typical need for many vehicle crossing events to be evaluated, computationally expensive structural analysis methods that require much computation at each time step (e.g. matrix inversion) are possible but not preferred. Consequently, a consensus has emerged around the use of influence lines, influence surfaces, and lane factors, that minimizes computation wherever possible, allowing very long simulations of traffic streams (e.g. 10,000 years) to be feasible with desktop computers. Thus, the static LEs due to traffic on a bridge deck are typically obtained from a series of simple (static) analyses of the different static systems. In this section, these methods are outlined, and recommendations given to ensure accurate evaluations.

3.4.1 Influence lines

Influence lines are frequently used in traffic-loading studies. They give reasonably good estimates of the longitudinal global LEs, and can be adapted, as explained later, to account for lateral distribution of load. Typical representative LEs of interest include maximum positive bending moment in a simply supported bridge, shear force at the support or negative support moment at the center of a 2-span bridge. An important suite of influence lines, typical of those used in traffic LE studies, is given in Table 3.2. As indicated, influence lines with positive and negative lobes can be truncated so that only the relevant loaded length is considered; that is, the maximum positive or negative effect. This is typically done for the development of a load model – Section 3.7 – but for determining the actual effects, the complete influence line should be used (Guo & Caprani 2019).

In the calculation, the train of vehicles, represented by point loads, is passed over the one-dimensional bridge in small steps, e.g. 0.5 m or 0.2 m (see Section 3.4.3). Figure 3.6 shows the LE calculation for a given arrangement of vehicle (or vehicles) on the bridge at a given point in time. The procedure for calculating the LEs due to a train of vehicles is the same as for a single vehicle – all the axle weights are applied as point loads on the influence line. The LEs due to each axle are calculated and combined by means of superposition to give the total effect. Moving the vehicle(s) to the next position in predefined increments creates the LE history, examples of which are shown in Figure 3.7.

3.4.2 Influence surfaces

Influence lines (e.g. Table 3.2) give LE as a function of the unit load's longitudinal position along the length of the bridge. By only considering

Table 3.2 Influence Lines Used in Background Studies to the Eurocode
(Bruls et al. 1996)

Ref.	Description of the Influence Line	Shape of the Influence Line
I1	Maximum bending moment at mid-span of a simply supported beam	
I2	Maximum bending moment at mid-span of a double fixed beam with an inertia that strongly varies between mid-span and the ends	
I3	Maximum bending moment on support of the former double fixed beam	
I4	Minimum shear force at mid-span of a simply supported beam	
I5	Maximum shear force at mid-span of a simply supported beam	
I6	Total load	
I7	Minimum bending moment at mid-span of the first of the two spans of a continuous beam (the second span only is loaded)	
I8	Maximum bending moment at mid-span of the first span of the former continuous beam	
I9	Bending moment on central support of the former continuous beam	

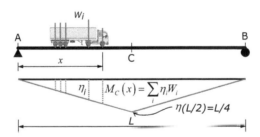

Figure 3.6 Calculation of mid-span bending moment, M_c in a simply supported beam under a truck load using its influence line. The parameter, $\eta_i(x_i)$ is the influence line ordinate corresponding to the ith axle at location x_i.

longitudinal position, these functions do not take account of the vehicle's transverse position. Extending the influence line concept into the transverse direction as well, gives the influence surface, which is defined as the LE at a point as a function of both the longitudinal and transverse positions of the applied load. With an influence surface, a vehicle can be positioned at any point on the bridge, and the LE resulting from each individual wheel load can be found and added. This approach is applicable where transverse positioning of vehicles within their lanes is critical for obtaining accurate results. An influence surface rarely has a closed-form expression, and linear

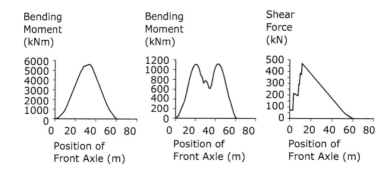

Figure 3.7 Responses of a 50 m long bridge to a single 5-axle truck for (left-to-right): mid-span bending moment in a simply-supported bridge; hogging moment at the central support of a two-span bridge; left hand support shear force in a simply-supported bridge.

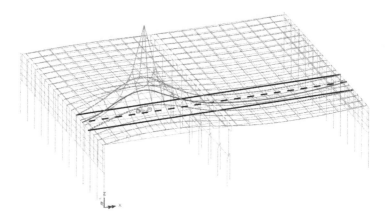

Figure 3.8 An influence surface for mid-span bending moment in a girder of a two-span integral bridge, showing wheel paths as slices and a 'mean' lane influence lane.

or quadratic interpolation between adjacent points (where the surface is defined) is required to determine influence ordinates for arbitrary positions of wheel loads. Because of this need for interpolation, it is best to use as few points to describe the influence surface as possible, to reduce look-up computational time. For example, planar portions of influence surfaces can be described using only four points, regardless of the dimension. Figure 3.8 shows a typical influence surface that is obtained from a grillage model and so is only defined at discrete points on the bridge deck, resulting in the need for interpolation for intermediate positions of axles.

A second approach, that has lower fidelity but requires less computation, is to use slices or cross-sections through the influence surface corresponding to the wheel paths of the vehicles. Assuming that the vehicles run perfectly along the center line of the lane, and reasonably assuming that wheel loads are equal, then the mean of the influence ordinates for the two wheel paths through the influence surface produces an influence line representing the lane. Referring to Figure 3.8, this lane 'mean' influence line can be seen along the center line of the lane and is the weighted average of the two influence lines of the wheel paths. It is important to note that the mean slice is not necessarily the same as a slice through the influence surface at the center of the lane, since the influence surface may have a highly nonlinear shape transversely (as is the case in Figure 3.8). For each lane on the bridge, this process is followed, and the total LE can then be summed over all lane influence lines to obtain the total LE at the point of interest. Typically, for this method, the lane influence line is only determined numerically (e.g. Caprani et al. 2012), and an interpolation between adjacent node points is needed for arbitrary positions of load, which again suggests the use of as few points as possible to reduce look-up times in the vector of influence ordinates.

A third approach for accommodating transverse load distribution is far less computationally demanding and is sufficiently accurate where the influence surface is approximately linear in the transverse direction. In this case, the influence surface is represented by a single influence line shape, but with a weighting factor for each lane for scaling. Again, the total component LE is found by summing the effects over all such lane-weighted influence lines. Although this approach assumes a linear transverse variation of the influence surface for the component of interest, this is quite reasonable for many situations. A further benefit of this method is that the influence lines can often be expressed as a closed-form expression, negating the need for additional interpolation computation.

3.4.3 Load movement

For accurate results, an important setting in the calculation of LEs is the step or increment size, Δx, by which the vehicle is moved across the bridge. Moreover, consider when there are multiple vehicles with differing speeds: in this case, vehicles are not incremented through a position change, but through a time step, δt, from which the increment in the vehicles' positions is determined from each vehicle's speed, v_i, as $\delta x_i = v_i \delta t$. It can be seen from Table 3.2 that the peaks of influence lines around the maxima are smoother in some cases than others. This gives an indication of the sensitivity of the calculated maximum response to the step size. Sharp peaks in the influence line manifest themselves in peaks in the responses as can be seen in Figure 3.7. In particular, the accurate determination of the maximum shear (or reaction) is clearly strongly dependent on the step size adopted, due to jumps in the response as axles enter the bridge. Other LEs can also exhibit this phenomenon (e.g. torsion at mid-span of a box-girder). The obvious solution is to use a very small step size, but this increases the computational effort. A further consideration is that it is quite usual that more than one LE (influence line) is being simultaneously calculated for the given traffic topology. Therefore, it is recommended to undertake an initial sensitivity study of the results to the chosen step size for the particular influence lines under consideration.

As an example, a sequence of vehicles on a 50 m long, two-lane bi-directional traffic bridge is simulated using a range of time steps, for the three LEs of Figure 3.7. The results are shown in Figure 3.9. For the LEs with influence lines that are continuous (and have no discontinuities), the responses are smooth and not very sensitive to the time step. For the influence line with a jump discontinuity (Load Effect 3), the response is seen to be sensitive to the time step adopted. For this example, Load Effect 3 is underestimated by nearly 2% for the 0.05 s time step and by nearly 10% for the 0.2 s time step. The other LEs are not significantly affected.

3.5 DYNAMIC INTERACTION

The preceding analysis has involved solely static calculations. As is seen within the analysis of Chapter 4, there are factors of vehicle-bridge dynamic interaction that are particularly significant for the extreme

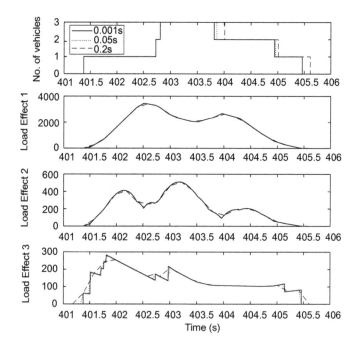

Figure 3.9 Static responses for three LEs using a range of time steps on a 50 m long bridge.

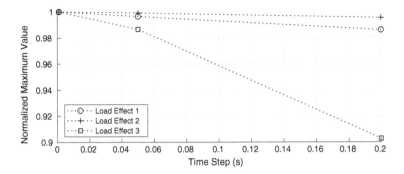

Figure 3.10 Error in static maximum response due to time step.

loading effects on short- to medium-span bridges. It is common practice that these static LEs are amplified to account for vibration effects. This is straightforward to do and to accommodate in the statistical prediction analysis that is to follow. However, in recent years an alternative refined treatment of dynamic effects has emerged which can be incorporated into the extrapolation of LE.

Firstly, it is mostly well established that for typical highway vehicles the dynamic increment reduces with increasing static LE; essentially, heavy trucks do not 'bounce around' as much as lighter trucks, even though the total stress is greater. In single degree of freedom systems, as the mass increases, the amplitude and frequency of the response decreases. Secondly, the dynamic amplification of static LEs under multiple trucks is even lower (Caprani 2005, González et al. 2008, OBrien et al. 2009, Caprani et al. 2012). Indeed, this is reasonable considering the destructive interference that occurs between the multiple vehicle dynamic systems and the bridge system. Together these vibratory effects mean that at increasing static values of LE, and thus reduced probability of occurrence, the dynamic allowance reduces. A final complexity of the dynamic effect is its intrinsic uncertainty: even for the same vehicle(s) and bridge, the system is sufficiently complex that it is not deterministic. Thus, this uncertainty exists alongside the reducing dynamic increment with reducing probability of occurrence, and both can be captured through a probabilistic consideration of total, static plus dynamic, load effect.

3.6 STATISTICAL PREDICTION

3.6.1 Prediction of extremes

Extreme low probabilities of exceedance are often expressed in terms of the average recurrence interval, more commonly known in engineering as the return period. Codes of practice often express design loading through statements or definitions such as that with a probability of exceedance of 10% in 100 years. The return period, R, is related to a design life, T, and a probability of exceedance, p, as follows (Ang & Tang 2007):

$$R = \frac{1}{1-(1-p)^{1/T}} \approx \frac{T}{p} \qquad \text{(Eq. 3.8)}$$

The LE at the return period R, can be referred to as the return level or characteristic maximum load. It is important to note that the return period, R, should not be confused with the design life, T. In Europe for instance, bridges are designed for a service life of 100 years with the capacity to resist LEs corresponding to a return period of approximately 975 years (5% exceedance in 50 years), often approximated as 1000 years (50/0.05). Thus, they are designed to carry a characteristic load that would be expected to occur in only approximately 10% of bridges just once in their lifetimes. These values coupled with appropriate partial factors based on relevant target reliability indices β_t then correspond to the design point (ISO 2394: 2015).

The artificial simulation of traffic and LEs can in principle be run for so many simulations that it produces LEs for any period required (Enright & OBrien 2013). Of course, the results are only as good as the assumptions inherent in the vehicle simulations, but this 'long-run simulation' approach has the advantage that the statistical distribution of the LE does not have to be assumed. In the absence of such extensive simulations, it is necessary to employ statistical techniques to predict the LEs at the specified return periods on the basis of extrapolation from a limited number of recorded traffic LEs.

A train of vehicles generates a response history of LE. This includes the responses due to single or multiple vehicles and load gaps in the history when there is no vehicle on the bridge. Since we are typically interested in the largest (or smallest, which is easily dealt with as the largest negative of small data) LE in a specified return period, it is necessary to consider the distribution of maxima $F_N(x)$:

$$F_N(x) = F^N(x) \qquad \text{(Eq. 3.9)}$$

where N is the number of events in the required return period and $F(x)$ is a distribution of the observed LEs (all data). This form of extrapolation dramatically amplifies minor errors or poor fitting in the upper tail of the distribution (Caprani 2005). It is therefore more appropriate to utilize asymptotic models and estimate approximate distributions based on extreme data only (Coles 2001). In practice this means that extrapolations should not be based on the distribution of all LEs, but on the distribution of maximum LEs in a specified block of time. Distributions of maximum-per-day, maximum-per-working-day, maximum-per-month, and maximum-per-year have all been used in relevant kinds of studies, depending on the quantity of data available (Bailey 1996, Zhou 2013, Zhou et al. 2012).

The time blocks used as a basis for extrapolation should be statistically similar. The block size should be large enough to accommodate any minimum variations, and the measurement or simulation period should be extensive enough to capture long term variations. Flows vary yearly, seasonally, daily, and hourly. Respectively, these variations are governed by economic activity, seasonal production demands, non-workdays, and the daily cycles of traffic, such as peaks hours and nighttime. Against this background, a common assumption has been to represent 'economic days' as those relevant for bridge loading. This is why weekend days and holidays are often separated from working days. If using daily maxima, there are then around 250 economic days per year (five working days for 52 weeks in a year, minus about ten for public holidays). This however depends on the national regulations, as in some countries, heavy traffic is permitted on the weekends and public holidays which effectively results in longer periods; care must be taken here with respect to the local conditions. As an example, South Africa has no imposed restrictions on heavy traffic operation and therefore the full 365 days are utilized in the preliminary development of the national load model (van der Spuy & Lenner 2019). Finally, given the variation in traffic flow during a day in most sites, it is not recommended to use a block size of less than one day. Chapter 6 considers this and other technical issues in more detail.

While there are many approaches to extrapolation, and Chapter 6 examines a range, the focus here is on a basic extrapolation using the block maximum Extreme Value theory. Given a dataset of, for example, daily maximum bending moments, with perhaps 250 samples (one economic year), it is of interest to determine the bending moment corresponding to a 100 year return period, as may be typical of a reliability analysis to justify extending the life of an existing bridge. There are three classes of Extreme Value distributions that are relevant to this data set, the Gumbel (Type I), Fréchet (Type II), and Weibull (Type III) distributions. To avoid the need to decide between these, it is easier to fit the Generalized Extreme Value (GEV) distribution which incorporates all three. The cumulative distribution function of the GEV distribution is (Coles 2001):

$$G(z;\theta) = \exp\left\{-\left[1+\xi\left(\frac{z-\mu}{\sigma}\right)\right]^{-1/\xi}\right\}$$

(Eq. 3.10)

in which the parameter vector $\theta = (\mu,\sigma,\xi)$ contains the location, scale, and shape parameters, respectively. The shape parameter distinguishes between the three classes of Extreme Value distribution: for Gumbel, $\xi \to 0$, Fréchet $\xi > 0$, and Weibull $\xi < 0$. Note that some publications express Equation (3.10) differently, changing the meaning of the shape parameter sign.

The shape parameter has significance because the Weibull distribution is bounded (has an upper limit), while the Gumbel and Fréchet distributions are unbounded (Fréchet is a heavy tailed distribution). Figure 3.11 illustrates the three upper tail behaviors of the GEV distribution for a standardized LE (i.e. location shifted to zero and width scaled to unity). The probability density and cumulative distribution plots are quite familiar. The Gumbel probability paper plot is frequently used in Extreme Value statistics, as it emphasizes very small changes in probability near 1.0 in the upper tail. The ordinate is a rescaling of the cumulative probability, referred to as the Standard Extremal Variate (SEV), as follows:

$$\mathrm{SEV}(z) = -\log\Big[-\log\big(G(z)\big)\Big] \qquad\qquad \text{(Eq. 3.11)}$$

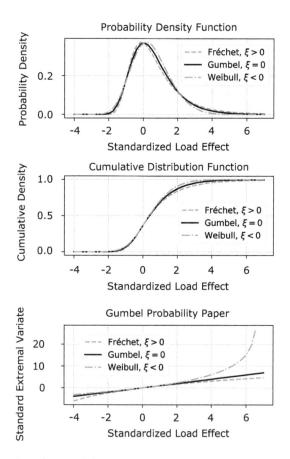

Figure 3.11 The three forms of GEV distribution (location = 0, scale = 1) with shape parameters of -0.15 (Weibull), 0 (Gumbel), and 0.15 (Fréchet), shown on three common forms of plot.

where 'log' is the natural logarithm. This rescaling of the vertical axis means that Gumbel distributions appear as straight lines, and so it is referred to as Gumbel Probability Paper (Ang & Tang 2007). Probability papers for other distributions are also sometimes used (e.g. Normal). However, it is recommended to prefer one form, so that the meaning and locations of curves in the plots of data become known with experience.

Noting the tail behaviors in Figure 3.11, it is frequently argued that bridge traffic loading is a Weibull-distributed process (or Gumbel at worst) (Bailey 1996, Caprani et al. 2003, OBrien et al. 2012, 2015). This is logical given that the Weibull distribution is bounded; i.e. it applies to data with a finite upper limit. It is reasonable to believe that traffic loading must have a physical upper bound corresponding to the maximum mass of manufactured vehicle types that can fit on a bridge. The Fréchet distribution is an unbounded distribution, and bridge traffic LEs should not fall within this distribution type. It is possible for bridge LEs to approach a Gumbel distribution, and several authors have applied it for this purpose. Yet, though the approach with Weibull seems to be favored, research by Nowak & Hong (1991) and Soriano et al. (2017) shows that fitting to a Normal distribution may work as well, but this has an unbounded tail like the Gumbel distribution.

Bridge traffic loading is a complex statistical process, with many subpopulations of randomness. For example, bending moment under 2-axle trucks will be distributed differently to those under 5-axle semi-trailers. Similarly, single truck loading events are differently distributed to events comprised of multiple trucks. These mixtures of data-generating processes mean that the use of a single probability distribution can frequently prove incapable of properly modeling the entire population of data. To deal with this added complexity, two approaches have emerged: (1) fitting the single distribution to the tail of the data only; (2) fitting multiple distributions to each separated sub-population and recombining. The next section examines this second approach in more detail, and so here the tail-fitting approach is briefly explained, though more detail is provided in Chapter 6.

For fitting single distributions to mixed data, the mixtures around the middle of the data are to be avoided, and the upper tail data is assumed to be comprised of a single generating mechanism representative of extreme LEs. The challenge then is the selection of an appropriate tail length as this may have a significant influence on the extrapolated value, especially when the fitted data is due to a mixture of events. Various tail thresholds have been investigated in previous studies, including the upper $\sqrt{n}$, $2\sqrt{n}$, $3\sqrt{n}$, and upper 5% and 30% of values (Castillo et al. 1996, Crespo-Minguillón & Casas 1997, Enright & OBrien 2013, Heitner et al. 2016, OBrien et al. 2012, O'Connor et al. 2009, Soriano et al. 2017, Zhou et al. 2012) with many authors favoring the length $2\sqrt{n}$ as initially proposed by Castillo (1988). In using a tail-fitting approach, judgement needs to be used to identify a tail that represents a reasonable quantity of data, but where there is a clear and consistent trend. This is consistent with OBrien et al. (2015) who conclude that the distribution chosen is less important than the quality of the fit in the tail. A comprehensive summary of different techniques for tail fitting is presented by van der Spuy & Lenner (2019).

To complete this basic introduction to prediction, a simple extrapolation is illustrated by example. Consider that a client wishes to keep an existing bridge operational for another ten years without major refurbishment. A characteristic LE is required for the assessment, here considered as a 100 year return period. The bridge is located at a site for which three months of traffic data was collected (Section 3.2). The traffic data was used to compute daily maximum LEs for five economic days per week, for 12 weeks, giving $n = 60$ daily maximum samples (Section 3.3). The data, z_i, is sorted in ascending order, $i = 1,...,n$, and the empirical probability of each data point is found from:

$$\hat{G}(z_i) = \frac{i}{n+1} \qquad \text{(Eq. 3.12)}$$

Using this along with Equation (3.11), the plotting position (ordinate) of each data point on Gumbel probability paper can be found. Prior to the widespread usage of desktop computers, a Gumbel distribution could then be fitted through these points as a straight line, and its parameters found (Ang & Tang 2007). Nowadays, statistical computation methods that do not rely on empirical formulas for ordinates are preferred, such as maximum likelihood estimation. These yield the parameter estimates for the GEV distribution that best describes the data, $\hat{\theta} = \left(\hat{\mu}, \hat{\sigma}, \hat{\xi}\right)$. Figure 3.12 shows the GEV distribution fitted to the data on Gumbel paper. Since we are looking for the largest LE that can occur in 250 economic days per year over the 100 year return period, while working with underlying daily

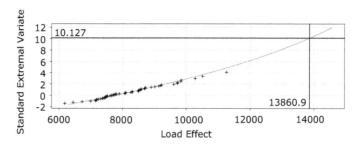

Figure 3.12 Example fit to sample data generated from a GEV (8000, 850, -0.15) distribution – 60 data points (daily maxima for three months) extrapolated to 100-year return period at 250 days per year.

maxima LEs, the probability that all LEs are less than this (i.e. the probability of non-exceedance) is:

$$p_{RL} = 1 - \frac{1}{250 \times 100} = 0.99996 \qquad \text{(Eq. 3.13)}$$

Using this, and given the parameter estimates, the return level at the 100 year return period is found from the inverse cumulative density function of the GEV, giving:

$$z_{R=100\text{yrs}} = G^{-1}\left(p_{RL}; \hat{\theta}\right) = 13861 \qquad \text{(Eq. 3.14)}$$

Also shown in Figure 3.12 is the return level SEV value, found from:

$$SEV_{R=100\text{yrs}} = -\log\left[-\log\left(p_{RL}\right)\right] = 10.127 \qquad \text{(Eq. 3.15)}$$

This provides a useful visual aid of the extrapolation 'distance.' Furthermore, it is at the intersection of this line, with the fitted distribution, that the return level is found, as shown.

3.6.2 Composite Distribution Statistics

In the previous section, it is noted that highway bridge loading events are known to arise from a mixed population (e.g. 1-truck and 2-truck loading events), and that this causes challenges with extrapolation. Caprani et al. (2008) introduce Composite Distribution Statistics (CDS) to account for the mixed population. CDS combines the relative contributions of different types of loading event to the overall probability of a specified level of LE not being exceeded – the non-exceedance probability. This is an interesting alternative to the conventional approach of tail-fitting a single distribution, as it provides an insight into the types of loading events that govern the (unobserved) extreme values of LE. CDS takes each distribution separately and combines the results by calculating the probability of non-exceedance of a specified LE level as the product of the component non-exceedance probabilities. The total probability of non-exceedance is the probability of not being exceeded by 1-truck loading events, and not being exceeded by 2-truck loading events, and so on. As the event type occurrences are independent, these component probabilities are multiplied, and the composite distribution of non-exceedance is:

$$G_C(x) = \prod_{i=1}^{N} G_i(x) \qquad \text{(Eq. 3.16)}$$

where Π indicates product and $G_i(\cdot)$ is the cumulative distribution function for the i-truck loading event of the N loading event types.

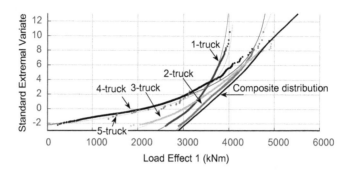

Figure 3.13 Daily-maximum LE distribution fits for *i*-truck free-flow traffic-loading events and the overall composite distribution for mid-span moment in a 40-m simply supported bridge.

Figure 3.13 illustrates an example of the CDS method on Gumbel probability paper. Noting that the increasing ordinate corresponds to increasing rarity, it can be seen that 2-truck loading events appear as the critical form of loading up to a LE level of around 4400 kNm, where 3-truck loading events become more critical. The total probability of LE exceeding a specified level includes all the component probabilities. It is particularly important to note that a short period of observation (e.g. months), corresponding to lower values of SEV, suggests that 2-truck events are critical, whereas longer observations (many years) actually reveal that the rarer 3-truck events are more critical (at SEV around 8). For example, at the 4400 kNm level, the contributions from 1-truck and 5-truck loading events are negligible. At a LE value of around 4800 kNm, the 3-truck event curve merges with the composite distribution, showing that all other event types have become negligible. This is despite the clear governance of 2-truck events around the mean (ordinate value 0.36; LE value around 3400 kNm). Indeed, this is quite typical: the 'average' block-maximum loading events are not necessarily reflective of the composition of the truly extreme loading events.

3.6.3 The governing form of traffic

The governing form of traffic is either free flowing traffic, with an allowance for dynamic amplification, or congested traffic, when vehicles are closer together but there are no dynamics. The governing form is generally considered to be a function of loaded length – the adverse portion of the influence line for the LE of interest. An indicative summary is given in Table 3.3. Beyond a certain loaded length, usually taken to be around 40–45 m, congested traffic is traditionally thought to govern. This 40–45 m length appears to have arisen based upon the values of dynamic amplification that have been so far included in codes of practice. Vehicles traveling in free flow at full highway speed are typically more than 25 m apart. Vehicles in

Table 3.3 Governing Forms of Loading for Different (Approximate) Loaded Lengths

Approximate Loaded Length	Governing Case	Traffic
Point locations (e.g. expansion joints), or lengths < ~1.0 m	Single axles	Free flowing
Less than ~8.0 m	Axle groups (tandem, tridem, etc.)	Free flowing
~8.0 to ~20.0 m	Single vehicle per lane	Free flowing
~20.0 to ~40.0 m*	Multiple vehicles per lane	Free flowing
> ~40 m*	Multiple vehicles per lane	Congested

*The point where congested traffic becomes more onerous than free flowing has been the subject of much debate.

stopped traffic can be as little as 2 m apart. Hence, although no dynamic allowance applies, LEs from congested traffic tend to be greater on longer loaded lengths in comparison to the free-flowing traffic with a dynamic allowance. In the past few decades, a number of factors have emerged to challenge the perception that congested takes over from free-flowing traffic at around 40 m, exposing a deeper complexity to this issue. The value taken for the dynamic amplification of static LEs is a key consideration. While values of 1.2 and 1.4 have been used in the derivation of the Eurocode load model, much lower values have been suggested in more recent studies, as is discussed in Chapter 4. When the allowance for dynamics is indeed less than previously used, then the LEs due to congested loading become critical at much lower loaded lengths.

Using CDS, Caprani (2013) proposes a statistical model to capture the preceding complexities, and gives the total distribution of LE as a combination of dynamic, free-flow, and congested components:

$$G_T(z) = \left[D(\cdot)\, G_C^{FF}(z) \right] G_C^{CF}(z) \qquad \text{(Eq. 3.17)}$$

In this formulation, the composite distributions of free-flow ('FF') and congested flow ('CF') are found using the usual traffic streams, LE calculations, and statistical estimation methods that have been described above. To account for dynamic effects, Equation (3.17) includes a purposefully vaguely defined 'dynamic function,' $D(\cdot)$. This dynamic function is defined by the user, and can arise from site measurements, a code of practice, or advanced dynamic and probabilistic modeling. It could be a constant (e.g. $1.3\,z$ as defined in Australia), or reduce with increasing load effect, z, or it could even depend on the probability of the load effect, z.

An important factor when combining traffic states is that the relative periods of congestion and free-flow traffic at the site should be considered. For example, if it is considered that the site will experience two hours of congestion per day, then 200 hours of simulated congested traffic and 2200 hours of free-flowing traffic will correspond to 100 days; the daily maxima

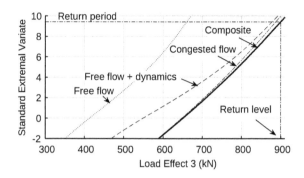

Figure 3.14 Extrapolation for LE distributions of free-flow, free-flow plus dynamics, congested flow, and the composite, or overall, distribution, for a site with 0%
cars using the AASHTO LRFD DAF model (shear in 40 m simply-supported
span).

for free-flow and congestion is then taken as the maxima of the blocks of
22 and 2 hours, respectively.

Caprani (2012) illustrates an example of this approach in a probability
paper plot, reproduced here as Figure 3.14. At the return level, no single
form of traffic governs, but a mixture of both free flow (including dynamics) and congestion actually governs. Analysis using this total site LE framework suggests that, for loaded lengths where neither free flow nor congested
traffic dominate, both should be considered. The frequency of occurrence
of each traffic type is also significant. However, what is clear from the
results of this approach so far, is that congested traffic governs at much
shorter spans than previously considered. A side effect of this is that bridge
lengths and LEs previously thought to be governed by free-flow traffic plus
dynamics, could be found to be governed by congested traffic, at a slightly
lower level of characteristic LE.

3.7 NOTIONAL LOAD MODELS

3.7.1 General method

As outlined in Chapter 1, a Notional Load Model (NLM) is a set of loads
that replicate the extreme effects of real traffic streams when applied to
a structure. A well-developed notional load model balances the need for
simplicity with an attempt to provide close approximations of target characteristic LEs for the range of spans and influence lines considered. The
load model should perform well at the critical scenario, even though this
may result in excess safety for other situations. While there is an inevitable conservatism tied to this, at the design stage the cost of investment is
offset by the additional safety margin gained by some LEs and spans and

in-built allowance for future traffic growth in those cases. In contrast, for the assessment of existing structures, knowledge of the actual traffic loads should be exploited wherever possible. Indeed, NLMs can be developed for a single bridge but are more commonly developed to apply to a larger number of bridges across a region or specific road network. As discussed in Section 3.1, the design of an NLM is typically based on relevant traffic data. This section outlines the process of moving from the traffic data to the NLM itself.

The target values of characteristic LE that the NLM should replicate must be determined. To do this, the range of applicability of the model should be defined. This will determine the traffic data that is required, and influence lines for which the characteristic LEs should be determined. For example, if the NLM is for a single bridge – sometimes known as a Bridge Specific Allowable Load Limit (BSALL) – then the influence lines will be those for that bridge. On the other hand (and more commonly), when the NLM is aimed at all of the bridges over a large geographic region, a suite of influence lines, span lengths, and the most onerous traffic data for that region should be determined. For example, in the development of the Eurocode, the model was based on traffic data from the heavily trafficked Auxerre site and span lengths of $L = 5$, 10, 20, 50, 100, and 200 m were considered for each of the influence lines given in Table 3.2 (Bruls et al. 1996). Following the procedures outlined in the previous sections, the target characteristic values of LE are then found for each span length and influence line.

Notional Load Models are defined to replicate the target characteristic values of LE across all span lengths and influence lines considered. Almost all NLMs use a combination of:

- Point loads to replicate shear LEs;
- A uniformly distributed load (UDL) to replicate bending moments.

With this configuration, the challenge is then to optimize the values of the point loads and UDL to envelope the target characteristic LEs. This is usually done through ad hoc and iterative approaches, and commonly takes into account the pre-existing load model configuration in the jurisdiction.

In selecting the point loads of the NLM, it is interesting to note that while the primary purpose of an NLM is not to represent a typical vehicle, but to replicate the characteristic LEs for a wide range of bridge lengths and influence lines, many code developers do use realistic vehicle silhouettes (e.g. the US's HL-93, Australia's SM1600, China's D60). Since, of course, the NLM will prescribe very high values of loading, it can be argued that realistic vehicle silhouettes can mislead road freight operators in the allowable forms of truck traffic on the road network. More abstract NLMs, such as the Eurocode's 2-axle bogey or the older BS5400 single-axle knife-edge load, do not have this risk of misleading road freight operators. However, a controlling factor in the selection of the number and distribution of the point loads can be the historical legacy of previous design codes and what practitioners in the region are familiar with.

The UDL of the load model is found to reduce with increasing span. This is because the probability of many trucks exceeding a specified weight simultaneously in the same lane reduces as the number of trucks increases. Hence, for long convoys of trucks, the average weight of each truck tends to reduce with increasing convoy length. Some standards allow for this with a uniform loading of variable intensity. For example, the intensity of the uniform loading in the South African TMH-7 NA load model is illustrated in Figure 3.15. While such specification of loading is more accurate for longer loaded lengths, it is more complex to use (Lenner et al. 2017). As a result, most standards, such as AASHTO and the Eurocode, use a uniform loading of fixed intensity which tends to be conservative for longer spans.

For sites with multiple lanes, it is typical to initially develop the NLM for the lane corresponding to the heaviest traffic (Bruls et al. 1996). For the remaining lanes, the NLM point load and UDL values tend to be reduced with increasing number of lanes due to the reduced probability of multiple heavy vehicles simultaneously occupying adjacent lanes (van der Spuy et al. 2019).

3.7.2 Specific considerations

3.7.2.1 Design versus assessment

To develop a design standard for new bridges, a notional load model is generally specified which gives conservative estimates of characteristic maximum LEs for a range of bridge types and spans. The reliability calibration

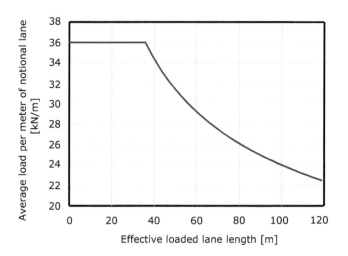

Figure 3.15 UDL as a function of loaded lane length (CSRA 1991).

of the load model seeks to provide a minimum required performance over the lifetime of the structure by specifying a target reliability value β_T. This results in partial factors for the considered limit state that are applied to the derived characteristic load (Steenbergen & Vrouwenvelder 2010). The safety calibration of the design load model is intended to be optimized for the entire range of structures (Rackwitz 2000); it is therefore inevitable that some bridges will be more conservatively designed than others. From the optimization perspective, it is wasteful to have a load model that results in widely different levels of conservatism for each LE. Care must be therefore taken in the specification of the span lengths and LEs, against which the model will be developed. For the assessment scenario the situation changes, as a specific bridge is considered as opposed to the entire stock. The characteristic maximum LEs can be factored and used directly in a bridge safety assessment (Pérez Sifre & Lenner 2019). In developing a site-specific load model, the primary goal is not to be conservative but rather to minimize the range of conservatism for the LEs and spans considered, i.e. to provide a consistent level of safety for all LEs.

The American HL-93 notional load model is intended to generate the 75-year return period characteristic LE for 'normal vehicular use of the bridge' (AASHTO 2010). WIM data was used to calibrate the US design load model (Nowak 1999), and NCHRP Report 683 details a method to adjust the US HL-93 model using site-specific WIM data to obtain a site-specific load model for design or assessment. The ability to adjust a model is useful – some roads are subject to much heavier vehicles than others and their bridges need greater load carrying capacity.

O'Connor et al. (2001) describe the process used to derive the Eurocode load model using WIM data from France. As for HL-93, the Eurocode's Load Model 1 represents extremes of non-permit and routine permit vehicles. This standard also incorporates 'α-factors,' scaling factors that can be adjusted for countries or networks (Marková 2013). Work towards a European-wide harmonized assessment load model is ongoing at the time of writing.

3.7.2.2 Load model consistency

As noted earlier, there is a tradeoff between the consistency of load model accuracy and its simplicity. The calibration can be governed by a few critical cases (spans or influence lines). This means that LEs can be inconsistent over the range of span lengths and structure types. Consequently, the NLM LEs for some bridge lengths or influence lines will be significantly more conservative than others. As an example, the consistency of the HL-93 load model was investigated by Leahy et al. (2014) who plot the ratio of characteristic maximum LE to the corresponding AASHTO HL-93 value, using data from 17 American WIM sites – Figure 3.16. Sub-unity values for this ratio confirm that the standard is conservative for these sites. Clearly some

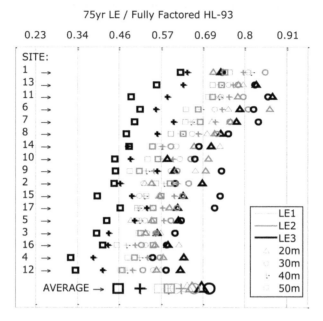

Figure 3.16 Single-lane characteristic LEs, normalized with respect to HL-93 (adapted from Leahy et al. 2014, with permission from ASCE). (LE1 is mid-span moment and LE2 is end shear in a simply supported bridge; LE3 is central support moment in a two-span continuous bridge.)

sites are much more lightly loaded than others, making the corresponding bridges safer. But the more interesting point, from the perspective of designing a load model, is the spread in the data for a given site. Three LEs were considered in this study for four different bridge lengths so twelve ratios of characteristic maximum to HL-93 LE are calculated in each case. It can be seen that the range of ratios is considerable. For Site 11, for example, the ratio ranges from about 0.5 to 0.87. A well-designed notional load model should seek to minimize this range, particularly for the more common bridge types and spans. This is often done in a load model by combining point loads and distributed loading to optimize the effect for both shear and bending moment. The need for point loads is readily observable when attempting to replicate the shear envelope, while the uniformly distributed load is particularly helpful in matching the bending moments. This results in a combination of point and distributed loads. As the matching exercise rarely results in a unique solution, care must be taken in the optimization of the desired load model.

3.7.2.3 Multi-lane factors

For multi-lane traffic, the critical loading event for a short to medium span bridge can involve extreme vehicles in one or more lanes. The main factor determining which case governs for the element under consideration is the transverse stiffness of the bridge. For example, a solid deck has a relatively deep slab and correspondingly high transverse stiffness. This results in a high degree of load sharing between lanes and a greater contribution to LEs from traffic in lanes remote from the location under investigation. It follows that maximum LEs for design tend to be due to adjacent heavy vehicles in multiple lanes rather than a single extreme vehicle. A beam-and-slab bridge, on the other hand, with only a thin slab of low stiffness connecting the beams, exhibits limited lateral distribution. This results in much lower load sharing between lanes and the governing load case is more likely to involve a single extreme vehicle.

There have been many studies on lateral (or sometimes girder) distribution factors that quantify the contribution of loading in a lane to the LE under consideration. The objective is to quantify LEs using a simple beam analysis (or influence line) for the most critical lane and scaling this result to consider the traffic in other lanes. A simple approach defines a component distribution factor as the LE caused by traffic in a given lane, expressed as a portion of the total (maximum) LE value. The components for each lane should sum to unity. Component distribution factors can be found by passing a vehicle or axle loading along each lane and using a Finite Element model to calculate the resulting LE. The resulting factors will depend on the vehicle chosen for the study but are not particularly sensitive to it. This approach is effectively assuming that the influence line for any specified lane can be found by scaling the influence line for the most critical lane.

An interesting distribution factor approach is the S/D concept developed for the AASHTO code, evolving from Newmark (1948) through Sanders & Elleby (1970) to Zokaie et al. (1991). Despite the range of approaches to lateral distribution factors, the essential idea is to eliminate the need to use an influence surface (or Grillage/Finite Element model) to determine LE.

Zhou et al. (2018) describe some flaws with prevalent methods and present a formal general model for the problem, using a multivariate Extreme Value approach to quantify the lane factors appropriate for a given traffic stream, LE, transverse distribution, and return period. This approach uses copulas to model the correlations at different probabilities between the different marginal distributions of each lane. The approach is complex, but rational, and yields site-specific lane factors which are less conservative than those in codes, and so suitable for important bridges.

To avoid the necessity to evaluate transverse stiffness and load sharing between structural components in specifying a load model, an approach based on multiple lane WIM data was developed to allow for considerations of extreme multi-vehicle events of desired levels of non-exceedance (van der Spuy et al. 2019). Permutations of LEs due to recorded WIM data, allowing observation of side by side events in any number of lanes, provide a basis for extrapolation and specification of multiple lane presence factors; an approach suited for the derivation of load models for design.

As discussed, notional load models used in standards typically use a combination of a uniformly distributed load and a set of point loads. Some standards use an arbitrary set of point loads, others prefer a notional vehicle, bogie, or tandem. The most critical or principal lane for the specified LE is subject to a lane load model. Portions of that principal lane loading are typically applied in other lanes to allow for the possibility of an extreme multi-lane loading event. These factors are distinct from the component distribution factors mentioned previously. But in essence, the load model derivation for loading on more than a single lane is achieved by the modification of the notional load model specified for the heavy slow lane.

3.8 RECOMMENDATIONS

Load model development for short- to medium-span bridges can be summarized by some essential points. It is first necessary to establish the composition and statistical description of the traffic currently operating on the roads, where the most important parameters include the frequency distribution of gross vehicle weights, axle loads, truck geometry, and spacings between individual vehicles. Typical means of collecting the required input is the data from WIM systems. An alternative approach is the identification of a governing event in the form of a critical truck crossing during the proposed service life of a bridge – the critical truck then represents a vehicle with a geometry and loads to cause a critical loading in the considered period. Most codes are however currently reliant on the collected WIM data which is subsequently cleaned and calibrated.

As the obtained WIM data contains information on the truck geometry and individual axle loads, beyond direct calculation of LE, this enables long-run simulation: a sampling and generation of artificial traffic events at long periods that can equal or exceed the specified return period. The often-highlighted advantage of this approach, coupled perhaps with scenario simulation, is the generation of events not observed at the WIM site. From either measured or artificial traffic, LEs can be calculated using the influence lines or surfaces for the different bridge types under consideration. Subsequent fitting of appropriate statistical distribution functions, often from the Extreme Value family, and following the statistical extrapolation of obtained LEs to the required return period, provides characteristic values representative of the return period. There is currently no consensus about the appropriate return period, and in absence of such, it is suggested to rely on nationally or internationally specified target reliability values that essentially specify the probability of non-exceedance that can be then tied to both design and characteristic values.

Key parameters to be observed in either technique are the inter-vehicle gap modeling and the governing type of traffic loading, where the type is largely driven by the selected dynamic amplification model. The importance of vehicle spacing is in capturing events involving multiple vehicles. At the same time, the transition from free-flow to congested traffic, with congestion defined by slow speed and reduced gaps between vehicles, coupled with the appropriate dynamic amplification of the free flow determines the governing form of traffic on bridges. Short spans are governed by single vehicle events and the dynamic amplification is dominant there, but medium span length bridges may be governed by a mix of vehicle events along with reduced dynamic interaction, and therefore care is warranted in deciding on the scenarios driving the development of the load model. The composite distribution approach points towards a careful consideration of the governing form of traffic through a cautious evaluation of both gaps and dynamic effects.

This chapter is mostly concerned with the development of static LE on the short- to medium-length bridges based on the collected data. Guidance is given for appropriate techniques of data collection, cleaning, and calculation of LEs. Either statistical extrapolations or numerical simulations are considered as suitable techniques for the determination of characteristic values for the load model calibration. The derivation of the load model itself is driven by the load patterns observed in the data. It is suggested to carefully consider both free-flow and congested traffic situations along with the epistemic degree of dynamic amplification for each, as this can largely influence the derived load model. For a complete derivation of a load model for short- to medium-span length bridges, factors such as return period, target reliability index, cost optimization of applied load, and model uncertainty need to be further investigated.

Chapter 4

Dynamic load allowance

Jennifer Keenahan, Eugene OBrien,
Aleš Žnidarič, and Jan Kalin

4.1 INTRODUCTION

4.1.1 The phenomenon

Bridges vibrate in response to vehicle-crossing events and this can result in an increase in traffic load effects (LEs). The phenomenon is illustrated in the very simple example of Figure 4.1. This shows the components of strain at the bottom of the beam in the center of a simply supported span, caused by the passage of a single point force. The static component is generally the largest, and in this case is triangular, peaking when the force passes mid-span (Figure 4.1(b)). The dynamic component, Figure 4.1(c), is approximated as a sinusoidally varying vibration at the bridge's first natural frequency. This example shows that dynamics can be adverse when this component has reached a peak in the time it takes for the load to reach the center of the bridge – Figures 4.1(c) and (d). Equally, the dynamic component may have reached a trough in which case, dynamic 'amplification' can reduce the total LE – Figure 4.1(e). Thus, the vehicle can be taken to have a 'pseudo-frequency' associated with its speed that can resonate or not with the bridge's first (or other relevant) natural frequency. If dynamic amplification is plotted against vehicle speed, a series of peaks can be seen, corresponding to this resonance effect.

When two point forces, corresponding to a 2-axle truck, cross a beam, the static peak at mid-span strain occurs when one of the forces passes mid-span. If the bridge is again assumed to vibrate at its first natural frequency, starting when the first axle arrives on the bridge, a similar situation occurs but there are two possible resonance events – a peak in bridge vibration could coincide with the arrival of either axle at mid-span. This theory was first proposed by Frýba (1996), a pioneer in the field, and the resonance effect was illustrated using simple models by Brady et al. (2006) and Brady & OBrien (2006). However, when the results of this simple theory are compared with field measurements, there is little agreement – real vehicle/bridge interactions are far more complex.

DOI: 10.1201/9780429318849-4

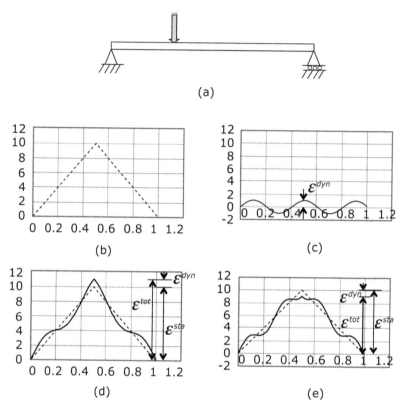

Figure 4.1 Mid-span strain due to a point force crossing a beam: (a) point force on bridge; (b) static response; (c) dynamic response for positive DAF; (d) total (static + dynamic) response for positive DAF; (e) total response for negative DAF.

In practice, a vehicle is not well represented by a series of point forces. A vehicle has mass and the movement of the mass will change the system's natural frequencies during the crossing event, especially for shorter bridges. The mass of a vehicle is supported on a number of springs and there are associated natural frequencies. Heavy vehicles generally have body bounce and pitch frequencies around 1.5 to 4 Hz and axle hop and suspension pitch frequencies around 8 to 15 Hz (Cebon 2000). The vehicle and the bridge interact dynamically during the crossing event and this interaction can be constructive or destructive. For example, if the bridge is moving downwards when an axle arrives, this movement tends to be amplified by the axle (constructive interference). If on the other hand, the bridge is moving upwards when an axle arrives, the vehicle and the bridge act against one another (destructive interference). All of this is complicated by the presence of the road (or rail) surface profile. The natural variations in a typical road surface are of the same order of magnitude as bridge deflections and cover the entire spectrum of frequencies. Hence, as a vehicle crosses a bridge, a wide range of motions are going on and several frequencies are at play.

OBrien et al. (2010) and Caprani et al. (2012a) propose a statistical approach to dynamics. They point out that it is not the ratio of total (static plus dynamic) LE to static LE for a single loading event that is relevant. Given that statistical studies of static loading are generally carried out first and an allowance for dynamics is applied afterwards, they suggest that a more relevant parameter is 'Assessment Dynamic Ratio' (ADR): the ratio of characteristic maximum total LE to characteristic maximum static LE. They calculate characteristic total LE by performing vehicle/bridge dynamic interaction analyses on large numbers of extreme loading events. The trend is quite interesting. While dynamic amplification for some non-critical loading events can be quite large, the trend amongst the extreme loading events is much more modest, with the calculated ADR typically not exceeding 10%. Znidaric (OBrien et al. 2013a) directly measures static and total LEs and provides further evidence that the influence of dynamics tends to diminish as static loading events become more extreme. This can be explained in part by the increasing complexity of extreme loading events, particularly those involving multiple vehicles, which increases the likelihood of destructive interference effects and tends to reduce the total LE.

Bridge loading standards generally make some allowance for dynamic amplification. This can depend on the type of load effect (LE), the span length, the number of lanes, etc. The notional load model for which bridges are designed, generally includes an allowance for dynamic amplification.

4.1.2 Basic definitions

As outlined above, the load effects that a vehicle generates in a bridge arise from a combination of the static weight of the vehicle and inertial loads due to dynamic interaction between the vehicle and the bridge. The consequent total load effect, ε_T, is typically larger than that from a wholly static analysis, ε_S. The dynamic amplification factor (DAF) represents the ratio of these two load effects and is defined as:

$$DAF = \frac{\varepsilon_T}{\varepsilon_S} \qquad \text{(Eq. 4.1)}$$

An alternative definition of the same concept, used by some authors (e.g. McLean and Marsh 1998) and codes of practice (AASHTO 2012, Standards Australia 2017, 2017a), is the Dynamic Load Allowance (DLA), Dynamic Increment (DI), or Impact Factor (IF), given in % by:

$$DLA = DI = IF = \left(\frac{\varepsilon_T - \varepsilon_S}{\varepsilon_S} \right) \times 100 \qquad \text{(Eq. 4.2)}$$

4.1.3 Factors influencing dynamic amplifications

As will be seen later, codes of practice take very different approaches to the consideration of dynamic effects. This is because of the complexity of the phenomenon and the vast array of factors that contribute to it. These are the main contributing factors, and each is explained in more detail below:

- The road surface;
- The bridge structure;
- The vehicle(s);
- The loading event;
- The structural response (LE) under consideration.

The road surface condition is consistently found to be a strong indicator for the level of dynamic interaction that occurs (Deng et al. 2015). Smooth road profiles produce lower DAFs than rough road profiles. Many authors have explored the relationship between road surface quality indices, such as the International Roughness Index (IRI), and the resulting DAFs (e.g. dynamic amplification estimator, bridge roughness index, etc.). However, due to the many other factors affecting DAF, no general relationship has been found. However, one certainty that can be stated is that regular pavement maintenance to ensure a smooth profile reduces the level of DAF, with resulting benefits for bridge maintenance needs. This is a cost-effective approach to increasing bridge safety. In doing so, particular attention should be paid to bumps near the start of the bridge.

The bridge structure itself has an important role in the level of DAF that occurs. There is a strong inverse relationship between bridge length and first natural frequency (e.g. Chan and O'Connor 1990). Hence one is often used as a proxy for the other, as will be seen across various codes of practice. The literature (see Deng et al. 2015, Li 2006) is consistent in suggesting that short bridges are subject to higher dynamic interactions than longer bridges. One exception is that when frequency matching occurs between the vehicle and bridge, a higher DAF occurs. Most research has been done on short to medium span bridges (around 4–8 Hz) due to their prevalence and higher dynamic interactions due to frequency matching. The very common I-girder bridge form seems to be the most studied (Deng et al. 2015).

The main indicator of the level of dynamic interaction is the level of static load carried by the girder: girders that are more highly loaded statically, tend to have lower DAFs. However, as with all vibration problems, increased damping usually leads to reduced response, and DAF is no different. Timber bridges (for example) are found to have lower DAF than steel bridges (for example). Interestingly, new materials, such as fiber-reinforced polymer (FRP), do not exhibit the expected trend, and further research is needed on these forms of construction (Deng et al. 2014). Finally, Rattigan (2007) studies the influence of pre-existing vibration of the bridge (e.g. from a preceding loading event) on the resulting DAF of a new loading event. In some cases, the dynamic increment can more than double, but this requires a very particular set of circumstances (inter-vehicle gap and damping).

The vehicle or vehicles comprising the loading event are significant influencers on the level of dynamic interaction. By far the most significant influence is the mass of the vehicle. In particular, the ratio of the vehicle to bridge mass determines the degree of interaction between the two of them. A high ratio changes the overall dynamic properties of the system, and also results in higher static stress. It is widely found that dynamic amplifications decrease with increasing weight of vehicles or static stress (see, e.g. Figure 4.14). Many authors have investigated the influence of vehicle weight or static stress on DAF: Hwang and Nowak (1991), Huang et al. (1993), Chang & Lee (1994), Abdel-Rohman & Al-Duaij (1996), Kim & Nowak (1997), DIVINE (1997), McLean & Marsh (1998), Laman et al. (1999), Broquet et al. (2004), SAMARIS (2006), Li et al. (2008). Different vehicle types also exhibit different responses. For example, Cantero et al. (2011) examine the DAFs for a range of bridges subject to heavy articulated 5-axle trucks or cranes, allowing for bumps at the start of the bridge and for meeting events. For shorter spans (< about 20 m) the articulated truck yields higher average DAFs than the cranes, but this difference is less evident for longer spans.

One of the most frequently studied issues is the effect of vehicle velocity on DAF, as there is the potential for frequency matching between the crossing pseudo-frequency (velocity/length) and bridge frequency. Indeed, it has been shown that very high levels of DAF can occur, but usually only at very high speeds. However, these studies are usually theoretical and simplified, and when more realistic simulations (or field tests) are conducted, these predicted high DAFs are difficult to recreate (Zhu & Law 2002). Vehicle accelerations and decelerations are also found to affect DAF, with increased DAFs found in some cases for high decelerations (Deng et al. 2015, Li 2006). Different truck suspension systems are known to influence the dynamic amplification. For example, the DIVINE (1997) project reports that air suspension causes less (5–10%) dynamic amplification of static wheel loads than spring-leaf suspensions (20–40%). Indeed, Harris et al. (2007) show how a bridge-friendly vehicle suspension system can be designed that minimizes dynamic amplification.

Despite the body of research, there is no clearly identified link between DAF and the number of axles (Deng et al. 2014). Presumably, this is because it is difficult to isolate the effect of axle number from other properties such

as vehicle weight. In both field trials and numerical simulations, it is important to have sufficient approach length of the vehicle before the bridge so
that initial excitation can take place. This is merely to ensure that the calculated or measured DAFs are realistic. Finally, the transverse location of
the vehicle in its lane affects DAF but only insofar as it affects the level of
static load in any particular girder or element due to the transverse stiffness
of the deck. Then, the reduction of DAF with increasing static load is again
observed, and vice versa (Huang et al. 1993, Deng et al. 2015).

The loading event is characterized by the number of vehicles and the
inter-vehicle spacing. The literature consistently shows that the DAF for
multiple truck presence tends to be lower than for single trucks (Deng et al.
2015, AustRoads 2003). This is related to the increased static load (and
consequent lower DAF) that has been noted previously and to the likelihood
of destructive interference. For example, González et al. (2011) find that
for two 5-axle trucks meeting on any bridge over 12 m, the largest DAF
for shear is 5% and for bending is 1%. Similarly, Rattigan uses a calibrated
finite element model of the 32 m long Mura River Bridge (SAMARIS 2006)
and finds that in the worst case, two 5-axle truck meeting events give a
DAF of 15%. In terms of the inter-load spacing, theoretical studies have
found that for some idealized situations, a higher DAF can result for multiple moving point loads at particular spacings, than for a single point load
(Li 2006). However, for more realistic scenarios, DAF is found to consistently reduce due to the presence of more 'load' on the structure.

The final major factor governing the DAF is the structural response or
LE considered. There are many LEs that may be of interest, such as bending
moment, shear force, torsion, deflection, or stress. The reported DAFs vary,
depending on the response selected, even for the same bridge and vehicle
(e.g. Huang 2008, González et al. 2011). Validation of simulation models
against field trials is commonly conducted, but these usually only report on
strain or deflection DAFs. Therefore, simulation models (validated as far as
possible) are often used for other effects. Interestingly, the critical location
for static LE is not necessarily that for critical total (static + dynamic) LE (Li
2006). Cantero et al. (2009) find that the maximum total bending moment
occurs away from the mid-span of a simply supported beam, and define a
'full' DAF (FDAF) as the ratio of the maximum total response anywhere
on the beam to the mid-span static bending moment. They find this to be
greater than DAF and greater than unity (unlike DAF). Huang (2008) finds
similar results for a curved box girder. In short, it matters where, on the
bridge, DAF is measured.

4.2 CODES

Codes of practice and standards internationally take different approaches
to the issue of dynamic vehicle/bridge interaction. In general, codes seek to

Table 4.1 Summary of Some Current International Codified Rules for Dynamic
Allowance (*DLA* = Dynamic Load Allowance, *L* = loaded length in m, *f* = frequency) (Deng et al. 2015)

Country/Region	Contributing Factors	Summary of Allowance
United States[1]	Bridge length	$DLA = 0.33$ of truck portion of load
Europe[2]	Bridge length, no. lanes	$DLA = 0.3 - 0.2\left(\frac{L}{50}\right)$ $L \leq 50$ m
		$DLA = 0.1$ $L > 50$ m
Australia[3]	Number of axles	$DLA = 0.1 - 0.4$
China (MTPRC)[4]	Frequency	$DLA = \begin{cases} 0.05 & f < 1.5 \text{ Hz} \\ 0.1767 \ln f - 0.0157 & 1.5 \leq f \leq 14 \text{ Hz} \\ 0.45 & f > 14 \text{ Hz} \end{cases}$
Japan[5]	Length + Material	$DLA = \dfrac{20}{50 + L}$
Canada (CHBDC)[6]	Number of axles	$DLA = 0.25 - 0.5$

[1]AASHTO 2012, [2]CEN 2003, [3]Standards Australia (2017), [4]MTPRC 1989, [5]JRA 1996, [6]CSA 2006.

simply the approach and limit the number of contributing factors considered. A selection of approaches is summarized in Table 4.1.

4.2.1 AASHTO

The dynamic load allowance in the AASHTO specifications up to 2007, called the Impact Factor, was specified as a fraction of the static live load and as a function of the loaded length:

$$DLA = 15.24 / (L + 38.1) < 30\%$$

where L = loaded length in meters. In the AASHTO LRFD Specifications, used since 2007, dynamic load allowance is also specified as a portion of the static load, but independent of loaded length. Since the 2012 edition, for the Strength Limit States, the DLA is specified as 33% of the effect of the design truck only, with no dynamic allowance added to the uniformly distributed portion of live load.

The design dynamic allowances are based on field measurements. These indicate that the dynamic component is largely independent of vehicle weight. As static load is directly related to vehicle weight, the DLA reduces as ever heavier vehicles are considered. DLA is additionally reduced for simultaneous occurrence of two or more trucks in adjacent lanes. The maximum observed DLA in the measurements on about 100 bridges in

Michigan was less than 20% for a single truck and less than 10% for two trucks side-by-side.

4.2.2 Eurocode

The Eurocode for bridge traffic loading, EN 1991-2:2003, has been adopted throughout the European Union and several other countries worldwide. Factors, sometimes called 'α-factors,' are allowed to vary between countries, regions, and road networks within countries. The α-factors and other issues specific to a particular country are specified in a national annex.

The allowance for vehicle/bridge dynamic interaction is not explicit in the Eurocode but an allowance was incorporated during the development of the standard. Thus the notional load model for normal traffic, Load Model 1, is deemed to represent a characteristic maximum 1000-year value, including an allowance for dynamics, as appropriate. These built-in allowances are presented in Figure 4.2. It can be seen that they depend on the bridge length, the load effect (bending moment or shear), and the number of lanes of traffic. In multi-lane traffic, there is an increased likelihood of destructive interference, which reduces the extreme response. Clearly, in the development of the standard, these dynamic allowances were only applied to extremes of free-flowing traffic. For extremes of jammed traffic, it is generally assumed that no allowance is required.

4.2.3 Australian Standard

Australian Standard, AS 5100, specifies dynamic load allowances for vehicles traversing Australian road bridges. Part 2 of the code (AS 5100.2:2017) is used for the design of new bridges, whereas Part 7 (AS 5100.7:2017) is for the assessment of existing bridges. For the design of new bridges, the code specifies a range of loading arrangements, referred to as SM1600 loads, as shown in Table 4.2. The most critical load will depend on the bridge loaded length. It specifies a single deterministic DLA for each of these design loads

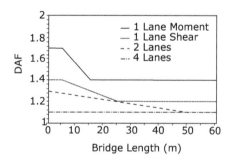

Figure 4.2 Eurocode allowances for dynamics.

Table 4.2 Dynamic Load Allowances for Australian Design Load Model, SM1600

Load Model Component	Number of Axles	DLA
W80 Wheel Load	1	0.4
A160 Axle Load	1	0.4
M1600 Moving Load	3	0.35
M1600 Moving Load	12	0.3
S1600 Stationary Load	12	0.0
Heavy load platform	16	0.1

that ranges from 0.0 to 0.4, depending on the number of axles acting on the bridge. Unlike most other international codes, load effect type, span length, and bridge frequency are not considered.

For bridge assessment, the DLA is specified as 0.4, with some exceptions. If, in the assessment, one of the SM1600 loads is selected as the traversing vehicle, then the value pertaining to that load (Table 4.2) prevails. Alternatively, if the road roughness is suitably low, a DLA of 0.3 may be adopted. Finally, the DLA can be reduced to 0.1 or even ignored (0.0), based on a reduced speed for the assessment vehicle.

The dynamic load allowances of AS 5100 were developed from a parametric study, considering single vehicle events on a small subset of representative Australian bridges (Austroads 2003). While it is acknowledged that multiple vehicle events are the critical load cases for many bridges, it is deemed appropriate to use a DLA based on a single vehicle, since the SM1600 loads are single vehicles. The study concludes that a relationship between dynamic load allowance and bridge frequency is present, but sees it as secondary to the influence of road roughness.

4.2.4 Chinese Standard

In the Chinese code for highway bridges (JTG D60 2015), the dynamic effect induced by vehicles on structural components is taken into account using a so-called impact coefficient, μ, whose definition is the same as the DLA.

In principle, the value of the impact coefficient is determined from the natural frequency of the structure, but other factors are also taken into account when determining the value, including the type of structural component and whether the LE is local or global. The design of superstructure elements in steel, reinforced concrete, and masonry arch bridges, together with supporting bearings and reinforced concrete abutments, should include the dynamic effect. Other bridge types are deemed to be unaffected by dynamics, including arches with pavement/fill thickness of over 0.5 m, culverts, and gravity abutments. The impact coefficient, as found from the impact coefficient formula, is shown in Figure 4.3.

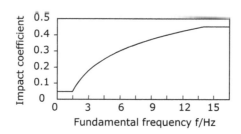

Figure 4.3 Impact coefficient (DLA) for China, as a function of fundamental frequency.

4.3 STATISTICAL APPROACH TO DYNAMICS

Most codes apply dynamic amplification factors to characteristic maximum static LEs or a notional load model deemed to represent these. If the worst static and dynamic cases are combined, such an approach is conservative because it does not recognize the reduced probability of two extremes (static and dynamic aspects) occurring simultaneously. The best example of this is that there are often very large dynamic amplifications for light trucks but much lesser ones for heavy trucks. Bridge-truck(s) interaction is sufficiently complex that the dynamic component of the LE may be considered as a random variable. Therefore, with any given crossing event, there are two resulting processes: static and total LE. For bridges, it is critical combinations of these two processes that are of interest.

Multivariate extreme value theory can be used to analyze critical combinations of several processes (Caprani 2005, González et al. 2008). Such an approach is more reasonable as it includes the respective probabilities of occurrence as well as any relationship between them. This theory is used here to incorporate the dynamic interaction of the bridge and trucks into an extreme value analysis for total LE. The results of this analysis can be used to determine Assessment Dynamic Ratio (ADR), a dynamic allowance that may be applied to the results of static simulations to determine an appropriate maximum total LE.

4.3.1 Assessment Dynamic Ratio

Assessment Dynamic Ratio (ADR) is defined as the ratio of characteristic maximum total load effect, $\hat{\varepsilon}_T$, to the characteristic maximum static value, $\hat{\varepsilon}_S$:

$$DAF = \frac{\hat{\varepsilon}_T}{\hat{\varepsilon}_S} \qquad \text{(Eq. 4.3)}$$

where $\hat{\varepsilon}_T$ includes both static and dynamic components of the LE. The key point here is that the characteristic maximum static and the characteristic maximum total do not, in general, correspond to the same loading events. In fact, they represent levels of the load effect rather than particular loading events.

The concept is illustrated in Figure 4.4 where each point is a maximum-per-100-year loading event. The maximum static and maximum total LE in the dataset are the same – point A – and DAF is the slope of a line from the origin to this point. Similarly, point B is the second greatest static and second greatest total LE. However, the third greatest static LE is point C whereas the third greatest total is point D. The corresponding ADR value is the slope of a line from the origin to point E and does not correspond to any particular loading event. ADR is, in effect, the ratio of the characteristic maximum total LE needed for design or assessment to the characteristic maximum static value, which can be found using the methods described in Chapter 3. In general, ADR is considerably less than the largest DAF values, which tend to be associated with less critical LEs such as point F.

To illustrate the ADR concept example, Caprani et al. (2012a) use the Slovenian Mura River Bridge and WIM data from the A6 motorway near Auxerre in France. A finite element model of the bridge is developed in which dynamic behavior of the model is calibrated against measured responses for single- and two-truck meeting events.

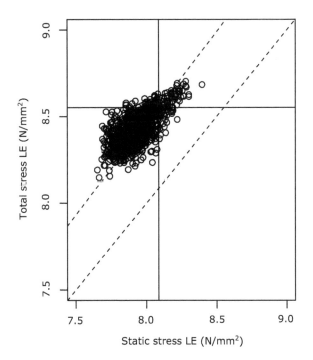

Figure 4.4 Maximum-per-100-year load effects (dN/mm²) (Caprani et al. 2012a).

For this site, ten years of bi-directional, free-flowing traffic data is generated numerically, and this traffic is passed over the influence line for one of
the girders to determine the static LEs that result. Each year of simulation
is broken down into 'months' of 25 working days each and there are thus
ten such months in each year of simulation (assuming 250 working days per
year). As a basis for further analysis, the events corresponding to monthly-
maximum static LE are retained. This is done to minimize the number of
events that are to be dynamically analyzed, as well as providing a shorter
extrapolation 'distance.'

The 100 monthly-maximum loading events obtained from the static simulations are analyzed using finite element bridge-truck interaction models.
Dynamic simulations are carried out and the results of the simulations are
a population of 100 monthly-maximum loading events for which both total
and static LEs are known. Scatter plots such as Figure 4.4 are drawn to
illustrate the results.

To capture the relationship between the static and total LE values, a bivariate extreme value distribution is fitted to the data, in this case a Gumbel
logistic bivariate distribution. The results are shown on the contour plot
of probability density in Figure 4.5. It can be seen that the most probable
results are around $(\varepsilon_S, \varepsilon_T) = (68, 71)$. The skew in the contours reflects the

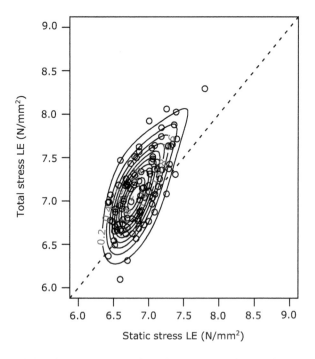

Figure 4.5 Probability density contours for a bivariate distribution fitted to maximum-
per-month data (Caprani et al. 2012a).

correlation between static and total – as static LE increases, total LE also tends to increase. As probabilities get less (outer contours), the correlation continues: for example, the last contour corresponds to maxima of $(\varepsilon_S, \varepsilon_T)$ = (76, 81). For most loading events, total exceeds static, i.e. DAF > 1.0, but this is not always the case, especially for less extreme loading events.

4.3.2 The shift in probability paper plots due to dynamics

OBrien et al. (2010) model the dynamic interaction between vehicles and a bridge using a 3-dimensional vehicle model traversing a finite element plate model, which takes into account the road surface roughness and characteristics of the truck fleet such as speed, weights, and suspension properties. Traffic simulations were used to generate 100,000 different annual maximum loading scenarios, 10,000 for each of five bridge lengths and two lane factors. Each of the 100,000 annual maximum events were analyzed using a vehicle-bridge interaction model for two ISO road classes ('A' and 'B') and two expansion joint conditions (healthy and damaged), adding up to a total of 400,000 dynamic analyses. The road profile and vehicle parameters were varied randomly within a Monte Carlo simulation scheme. The bi-directional traffic traversed the bridge, and the critical loading scenarios were found to be made up of anything from a single vehicle event to a combination of three vehicles.

The ADR values for 5, 50, 75, and 1000 year return periods are inferred by fitting the generalized extreme value distribution (GEV) to the top 30% tail of the annual maximum data. Figure 4.6 plots static and total bending moment for the 45 m bridge on Gumbel probability paper, together with the GEV tail fits. The studied return period levels of probability (5, 50, 75, and 1000 years) are shown. As can be seen, the distribution fits show excellent agreement with the data. The shift in the probability paper plot due to dynamics is clear, with the two curves being close to parallel. The ADR is the ratio of total LE to static, for a given level of probability. For example, for the 75-year return period, ADR is the ratio of bending moment corresponding to point B to that corresponding to point A.

The simulation results for mid-span bending moment in a range of spans are illustrated in Figure 4.7. The ADR values are quite small, with almost all results giving a value of less than 1.05. The road roughness has an effect, with Class B (ASTM 2015) road surfaces giving higher factors than Class A, typical of a well-maintained highway. The return period chosen has no significant effect. This is because the statistical trends of static and total LE are similar on probability paper (Figure 4.6), with the curves being roughly parallel. OBrien et al. (2010) use Finite Element analysis to derive two extremes in the lane factor to allow for the degree of load sharing between lanes. A 'high' lane factor of 1.0 is used to represent cases where the secondary lane is close to the beam of interest and/or the bridge is transversely

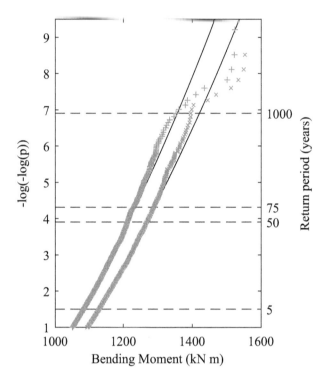

Figure 4.6 GEV fit of data, static (+), and total (×) bending moment on Gumbel probability paper, for 45 m span, high lane factor, Class 'B' profile (OBrien et al. 2010).

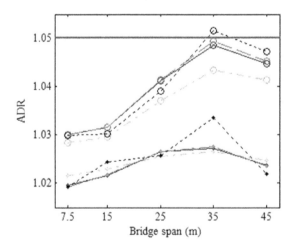

Figure 4.7 ADR values for 75- (_ _ _ _ _ .) and 1000- (...................) year return periods; for class 'A' (•) and class 'B' (○) profiles; (a) high lane factor; (b) low lane factor.

stiff. In this case, vehicles in both lanes contribute equally to the bending moment. A 'low' lane factor of 0.45 represents the case where the secondary lane is remote from the beam of interest and/or the bridge is transversely flexible. In this case, a vehicle on the secondary lane is deemed to contribute only 45% of its bending moment to the beam of interest. A comparison between Figures 4.7(a) and (b) does not show any significant difference in the ADR trends as a result of the lane factor assumed.

4.3.3 The contribution of surface roughness to dynamics

The condition of the road surface is another factor influencing the response of a bridge to a passing vehicle (DIVINE 1998) as well as specific locations of potholes and expansion joints (SAMARIS 2006). It is common to have a significant discontinuity at expansion joints as they are frequently the weak points of bridges and easily damaged (Lima & Brito 2009) or because differential settlements occur between the bridge and the abutment. A number of authors (Michaltsos 2000, Michaltsos et al. 2000, Li et al. 2006) show how a single pothole or bump, positioned at a critical location, can generate a large dynamic amplification. In particular, Chompooming & Yener (1995) show that combinations of pothole characteristics (i.e. height and length) and vehicle speed can result in very high dynamic effects. Clearly a bump in the right location can cause an axle or axle group to be in a 'down bounce' just when the static load effect is maximum.

OBrien et al. (2010) calculate Assessment Dynamic Ratio (ADR) for a wide range of spans, LE's, deck transverse stiffnesses, etc. Following a review of expansion joint surveys in Japan (Kim et al. 2007) and Portugal (Lima & Brito 2009), they considered the presence of a 20 mm deep depression over a 300 mm length, 500 mm before the center line of the bearing. Figure 4.8 shows the resulting 'Bump Dynamic Increment,' defined as the difference between the ADR calculated in the presence of the bump and the ADR calculated in its absence. Positive values indicate that ADR increases in the presence of the damaged expansion joint. It can be seen in the figure that the influence is only significant for the shorter spans and even then it does not exceed about 1.5%. For larger bridge spans, the bump makes little or no contribution. This is because the influence of the bump dissipates quickly and, for longer spans, its effect is negligible by the time the truck reaches a critical point on the bridge. Clearly bumps nearer to that critical truck location would be more significant.

4.4 FIELD MEASUREMENTS

Field measurement of dynamic amplification has clear advantages, as it takes account of all the uncertainties simultaneously – vehicle, bridge, and

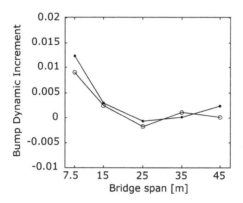

Figure 4.8 Bump Dynamic Increment for high lane factor and 50-year return period for class 'A' (•) and class 'B' (○) profiles (after OBrien et al. 2010).

road surface properties. A limitation is that it cannot provide the dynamic amplification appropriate for design directly, such as the ADR. In a number of cases, dynamic amplification of bridge response was measured for a finite number of known heavy vehicle crossings (Cantieni 1983, Deng et al. 2015). Measurement provides a good indication about bridge dynamic behavior under specific traffic loading and was used extensively in the past to account for the dynamic part of loading in bridge design codes. However, there are some drawbacks. The first of these is that each vehicle has specific dynamic characteristics depending at least on its type, gross weight, axle load distribution, axle configuration, and speed. This variability is a problem if the experiments only provide responses to a limited number of possible vehicles. Secondly, the vehicles used for such tests were typically not loaded above the legal limits. Thus, the dynamics of extreme illegally overloaded vehicles were not captured. Thirdly, responses to multiple truck presence events, and extreme overloaded vehicles are practically impossible to measure. This means that, without correction, field measurements tend to be biased towards levels applicable to common vehicle types and tend to give conservative values for DLAs. To collect reliable and useful information about the dynamic behavior of bridges under random traffic, it is desirable that the bridge dynamic response be measured for many thousands of random vehicles and multiple vehicle presence events. This is only practical if bridges are instrumented with sensors that measure the total response and the information is coupled with vehicle loading information, typically provided by a Weigh-in-Motion system.

4.4.1 History of using Bridge Weigh-in-Motion to estimate dynamic amplification

A Bridge Weigh-in-Motion (B-WIM) system uses an instrumented bridge as a weighing scale to estimate the static weights of passing vehicles. B-WIM systems can be used to find static and total responses simultaneously and hence to estimate the dynamic amplification. The first known attempt to measure the dynamic behavior of bridges with a B-WIM system was done by Ghosn & Xu (1989). They successfully show the potential of an extended B-WIM algorithm, on some bridge types, to estimate dynamic response under random traffic. A few years later Nassif & Nowak (1995, 1996) took more extensive measurements of dynamic behavior under random traffic. They coupled a dedicated acceleration measurement system with a commercially available B-WIM system to test four bridges with spans from 9 to 24 m and evaluated the dynamic amplification by comparing the dynamic signal with the static B-WIM approximation. The latter was obtained by filtering the dynamic signal with a Fast Fourier Transform. The filtering parameters were selected based on the experience of the researchers. The dynamic effect was described with a Dynamic Load Factor (Equation 4.1).

Nassif & Nowak (1995, 1996) approximated the static strain with a filtered total strain response. A few tens of DLFs were calculated on each bridge, for each individual girder, and were presented as a function of maximum static stress. Nassif & Nowak conclude that the dynamic component of stress or strain (i.e. the dynamic increment, ε_D) is practically independent of the static component and that, as a result, the DLF decreases with increasing static stress/strain. Furthermore, they note that for very heavy trucks, the DLF does not exceed the theoretical results. Two years later, Kirkegaard et al. (1997) came to similar conclusions by calculating the dynamic impact factors from selected simulated truck-crossing scenarios.

More extensive attempts to use B-WIM for the calculation of dynamic amplifications have been made possible with the further development of B-WIM algorithms, which can provide the dynamic amplification for loading events in real time (Žnidarič et al. 2008). In this study, the static response is estimated (dashed curve in Figure 4.9) as the sum of the contributions of each axle, calculated using the bridge influence line and the axle weights estimated by the B-WIM system (dotted curves in Figure 4.9). The disadvantage of this method was that any inaccuracy in the axle load estimates could have a significant influence on the accuracy of the static response

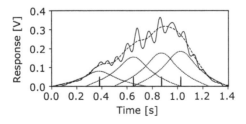

Figure 4.9 Measured response of a bridge to a 4-axle truck in volts (proportional to strain), including total (solid curve), estimated static (dashed), response to individual axles (dotted), and relative axle locations in time (solid spikes).

and hence on the inferred DAF. The two main reasons for errors in B-WIM axle load calculations are the ill-conditioning of the system of equations (OBrien et al. 2009a) and misidentification of axles (Žnidarič et al. 2017). Unfortunately, the likelihood of miscalculation of axle loads increases in cases where the DAF is likely to be significant, i.e. on bridges susceptible to dynamic excitation from traffic loading. Therefore, DAF values measured in this way, especially the higher values, have to be treated with caution.

In the *ARCHES* project (González et al. 2009), thousands of DAF values on several bridges were calculated and measured, through simulation and B-WIM measurement. A similar reduction in dynamic allowance with vehicle weight and corresponding load effects was observed as in Nassif & Nowak's (1995, 1996) study.

4.4.2 DAF results inferred from Bridge Weigh-in-Motion

In recent years, an alternative method has been developed to calculate DAF values with a B-WIM system. Similar to Nassif & Nowak (1995, 1996), it calculates the static approximation of bridge response by filtering the measured dynamic signal with a Fast Fourier Transform (Kalin et al. 2021). The significant advancement compared to previous attempts is that it defines the optimal filtering parameters from a few tens or hundreds of bridge responses to random passing traffic, and then calculates the DAF values automatically for all bridge crossing events. This approach removes the need for an expert who selected the filtering parameters based on personal experience.

A synthesis of the recent DAF/DLA results calculated from B-WIM measurements is shown in Table 4.3 (Kalin et al. 2021). It compares results from 12 Slovenian and 5 US bridges, with influence line lengths between 5.5 and 35.0 m. On sites SI04 and SI06, the two opposite lanes were treated separately and marked as 'a' and 'b' in the table. The bridge superstructure types were of two major groups: B&S – beams/girders with reinforced concrete (RC) deck – and RC slabs. Measurements took place from a few days to almost two years and resulted in a few hundred to almost 750,000 DAF results.

Results clearly show that DAF values vary considerably from one bridge to another. Mean DAF values range between 1.03 and 1.23, whereas the maximum DAFs do not fall below 1.14 and can be as high as 2.39. It is important to note that most of the maximum values occurred due to lighter vehicles. Only on bridges SI02 and SI03 were they caused by vehicles that were loaded close to, but still below the legal limits.

4.4.2.1 Examples of DAF calculation

Figure 4.10 shows the bridge response resulting in the highest DAF of 2.39, measured on any of the test sites. It was caused by a relatively light

Table 4.3 Summary of the Sites and Results of the Analysis

Site	Type	Length [m]	No. Vehicles	DAF Mean	σ	Max	at GVW [t]
SI01	Slab	7.4	202	1.23	0.14	1.96	13.7
SI02	Slab	10.5	4590	1.10	0.10	2.31	39.8
SI03	Slab	12.0	1850	1.11	0.08	1.80	38.8
SI04a	Slab	5.5	516	1.03	0.03	1.23	24.6
SI04b	Slab	5.5	318	1.04	0.03	1.14	8.2
SI05	Slab	6.8	617	1.08	0.05	1.40	5.1
SI06a	Slab	9.7	865	1.10	0.06	1.35	8.6
SI06b	Slab	9.7	474	1.19	0.12	1.48	15.7
SI07	Slab	6.6	746,594	1.09	0.04	1.39	17.9
SI08	B&S	34.4	432,307	1.06	0.05	2.39	22.5
SI09	B&S	35.0	402,595	1.02	0.03	1.56	5.3
SI10	B&S	25.8	289,533	1.05	0.03	1.55	10.0
US01	B&S	12.2	1979	1.05	0.04	1.34	8.3
US02	B&S	10.4	15,295	1.08	0.08	1.86	13.4
US03	B&S	11.0	7398	1.05	0.06	1.75	18.4
US04	B&S	10.4	1608	1.08	0.08	1.81	20.9
US05	B&S	25.0	25,219	1.17	0.13	2.37	11.7

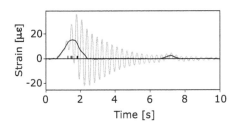

Figure 4.10 Event with the highest measured DAF among all vehicles from all sites.

20.4-tonne truck with axle groups 1-2-2 (FHWA Class 9), passing over the SI08 motorway test site. The resonance effects are clearly visible. The bridge oscillations are still present in the signal eight seconds after the vehicle has left the bridge, persisting through the passage of a 1.5-tonne car at around seven seconds.

Figure 4.11 shows responses for two vehicles crossing a bridge at site SI02. Figure 4.11(a) shows the highest DAF identified on this site. A 30-tonne 5-axle tractor semi-trailer with axle configuration 1-2-2 (FHWA Class 9) induces a DAF of 2.31. (It should be noted that the maximum total corresponds to the second tandem while the maximum static corresponds to the first tandem – an even higher DAF would have been recorded if the maximum static were taken from the second tandem peak). The resonance effect can clearly be seen as the trailer passes and may have been caused by the time between axle arrivals coming close to the first natural period of the bridge. The oscillation is sinusoidal around the static response, indicating a simple coupled motion. Figure 4.11(b) shows the bridge response to the passage of an empty 14.1-tonne tractor semi-trailer that resulted in a similarly large DAF of 2.26. In this case, the non-sinusoidal oscillation around the central values indicates a more complex vehicle-bridge interaction.

Figure 4.12 shows the response of the same bridge to the crossing of the heaviest recorded vehicle on this site. The DLA is less than 4%, much less than in the examples of Figure 4.11. This further supports the extensive evidence that heavier vehicles tend to generate less dynamic amplification.

There is further evidence of the trend of reducing DAF for heavier vehicles in site SI07. This site has almost two years of data, with around 750,000 recorded vehicle crossings retained for the DAF calculation, including a number of exceptional heavy transports. With only a 6.6 m long RC slab superstructure, this bridge was much less prone to dynamic excitation, which resulted in lower DAF values than on site SI02. Figure 4.13 shows the bridge responses for the crossing resulting in the highest DAF value and the crossing by the heaviest vehicle. The highest DAF of 1.29, shown in

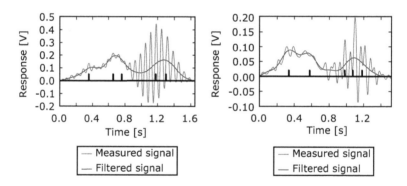

Figure 4.11 Vehicle crossings with the highest DAFs from site SI04: (a) vehicle A; (b) vehicle B.

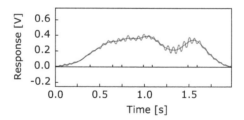

Figure 4.12 Event with highest GVW from site SI02.

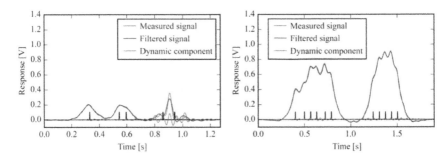

Figure 4.13 Extreme crossing events from site SI0I: (a) event with highest DAF; (b) event with highest GVW.

Figure 4.13(a), is caused by an empty tractor-trailer with axle configuration 1-2-2. In contrast, the permit vehicle travelling at 72 km/h, whose response is shown in Figure 4.13(b), is a 110.9-tonne low-loader with 11 axles that has a DAF of just 1.01.

4.4.2.2 Decrease in DAF with increasing GVW

One of the key findings from the literature and some recent codes is that DAF decreases with increasing load (Caprani 2013, González et al. 2009, Kirkegaard et al. 1997, Kalin et al. 2015 & 2021). To examine the evolution of DAF with increasing loading, Kalin et al. (2021) introduce a parameter, DAF_P, the mean DAF from a subset of vehicles, whose GVW is above a cut-off point at the Pth percentile of the GVW cumulative distribution. For example, DAF_{10} is the mean DAF of the heaviest 90% of vehicles, i.e. by discarding the lightest 10% of all vehicles. Similarly DAF_{50} is the mean DAF of those vehicles whose GVW is above the median weight. DAF_0 is thus the mean DAF of all the population.

The cut-off points, for the analysis, were chosen at the 0th, 10th, 20th ... 90th, plus the 95th, percentile. The value of DAF_{99}, calculated from the subset consisting of the heaviest 1% of vehicles, was calculated only for the four populations that numbered at least 100,000 vehicles. The grouping by percentiles of GVW, rather than by GVW themselves, allows the comparison of evolutions with GVW from all 17 datasets with substantially different GVW ranges.

As an example, Figure 4.14 shows the scatter graph of DAF against GVW for site SI10. Each point represents a DAF value corresponding to a loading event, either from a single or from several vehicles present on the bridge simultaneously. For multiple presence events, one DAF is calculated and is associated with each contributing vehicle's GVW in the figure. All permit vehicles with GVW above 80 t and most vehicles with GVW above 60 t have DAF values lower than the mean $DAF_0 = 1.05$. The heaviest exceptional

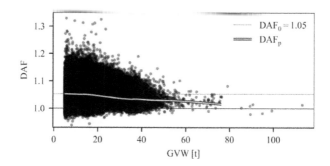

Figure 4.14 DAF values for site SI04.

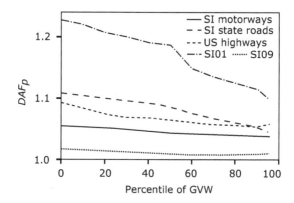

Figure 4.15 DAF_p values as a function of GVW percentile for three groups of bridges (SI = Slovenia, US = United States).

transport, weighing 113 t, had a DAF of only 1.013, making the dynamic increment of this extreme loading case similar to the load effect of a car. The decreasing DAF_p trace demonstrates the trend of decreasing mean DAF with increasing GVW.

Figure 4.15 summarizes the DAF_p results for all 17 datasets, grouped into three sets: Slovenian state roads and motorways and US highways. To allow for the datasets with different numbers of vehicles and different GVW distributions, the abscissa gives GVW in percentiles of the maximum. It can be seen for all three datasets that a) there are consistent decreasing trends of DAF with GVW and b) there is significant variation in the results, illustrated in the figure with the highest DAF_p values at Site SI01 and the lowest ones at Site SI09. In particular, Kalin et al. (2021) found that at the 35 tonne gross weight level, roughly matching fully-loaded large vehicles, either 5-axle semi-trailers in Europe or 18-wheelers in the US, the DAF_p values varied from 1.01 on the Slovenian motorway bridge SI09, to over 1.13 on the American highway bridge, US05.

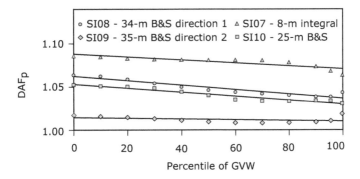

Figure 4.16 DAF$_p$ values for the four Slovenian motorway bridges.

To elaborate on the variability of the results, Figure 4.16 provides details on the four Slovenian motorway bridges. The results differ significantly, even between the two, almost identical, beam-and-slab twin bridges, SI08 and SI09. The measured first natural frequencies of both of these bridges were similar, 3.5 Hz and 3.3 Hz, and close to the typical whole vehicle body bounce frequency (Cebon 2000). The main difference was the expansion joint just before the instrumented span of SI08, which excited vehicles' body bounce modes of vibrations just before arrival on the bridge. Its twin bridge, SI09, is 0.6 m longer and structurally identical. However, with no expansion joint before the instrumented span, it does not exhibit similar resonance behavior and the DAF_p values are much less.

4.5 CONCLUSIONS

Design codes prescribe relatively large (conservative) dynamic allowance factors. Conservatism is not expensive for a new bridge and provides some extra reserves of capacity for the structure that may be useful in the later stages of its life. In contrast, for existing bridges, it can be important to obtain realistic measures of dynamics as input to a detailed traffic loading model; this can make the difference between rating the bridge as safe or unsafe. Further work is needed on methods of assessing dynamics in bridges but the evidence strongly suggests that dynamic effects in extreme cases of loading are quite small and much smaller than is allowed for in the major bridge standards.

Two main methods of measuring dynamics are described in this chapter, DAF and ADR. Of these, ADR is the more appropriate because, by definition, it is the ratio of what is required (maximum total LE) to that provided by a static load model (maximum static LE). However, deriving ADR is more complex, requiring a statistical analysis of the data. The concept of DAF is simpler and it has value but, given the evidence of reducing dynamics with increasing LE, it should be treated with caution and some allowance should be made for this reducing trend. Field measurements of DAF can provide valuable data but should also take account of the trend of reducing DAF with increasing vehicle weight.

Quantifying ADR or DAF remains a challenge. Simulations allow great quantities of scenarios to be considered but are only as good as the underlying assumptions on, for example, vehicle and bridge properties. Measurement can address this issue but there is a challenge with quantifying the static response. It is not practical to have vast quantities of vehicles and combinations of vehicles travel over the bridge at crawl speed. Recent Bridge Weigh-in-Motion data has provided large quantities of data but it should be treated with caution until further corroborating evidence becomes available to confirm the accuracy of the static estimates. Nevertheless, the data from Bridge WIM installations is extremely useful. It should be noted that Bridge WIM systems could also collect the data necessary to calculate ADR.

Chapter 5

Long-span bridge loading

Colin Caprani, Michael Quilligan, and Xin Ruan

5.1 INTRODUCTION

The critical loading events for long-span bridges are caused by congested traffic, see Figure 5.1. This is in contrast with short-span bridges where individual heavy vehicles or small clusters of vehicles in free-flowing traffic produce the maximum load effect (LE). During congestion events the gaps between vehicles reduce and the cumulative effect of closely spaced vehicles produces a greater LE, even though there is no amplification for dynamic effects because of the slow speeds involved. The key parameters that must be considered in the development of a long-span load model include the type and weight distribution of vehicles in the congestion event, their arrangement both in-lane and between lanes, their spatial distribution, as well as the duration and frequency of the congestion events themselves.

An important term in the study of long-span bridges, is 'loaded length.' Ivy et al. (1954) define this as follows:

> The 'loaded length' of a span is the length of loading required to produce the maximum live load stress in individual members or maximum reactions in supports. Loaded lengths for simple, continuous, or other types of structures are those lengths—or summations of lengths—producing the maximum stresses or reactions of the same sign in any member.

In this definition it is clear that the loaded length is different to the span of the bridge and depends on the shape of the influence line for the component under consideration. Almost all codes and studies refer to loaded lengths, rather than spans, when considering the design live load (Guo & Caprani 2019). Thus, different components of the long-span bridge will, in general, have different live load intensities for their design.

DOI: 10.1201/9780429318849-5

Figure 5.1 Traffic on Bosphorous Bridge, Istanbul.

5.2 LOAD MODELS FOR LONG-SPAN BRIDGES

Load models vary greatly between countries, but most codes specify a uniformly distributed loading, and this component tends to be critical for long-span bridges. Nowak et al. (2010) provide a comparison of the specified levels of uniform loading for several codes, shown in Figure 5.2. In most codes, the intensity of the uniform loading reduces with loaded length, reflecting the reduced probability of a very long convoy of heavily loaded/overloaded vehicles. In the calibration or development of these codes, it is usually assumed that the traffic is jammed, resulting in a stationary convoy of vehicles with small inter-vehicle gaps. The results are highly sensitive to the assumption on these minimum gaps (OBrien & Caprani 2005) but little statistical information on gaps in jammed traffic is available in the literature.

5.2.1 Development of North American load models

5.2.1.1 Ivy et al. (1954)

The first investigation into long-span bridge traffic loading in the United States was in 1953 with a study undertaken by Ivy et al. (1954). This

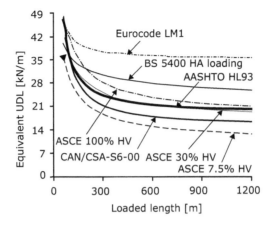

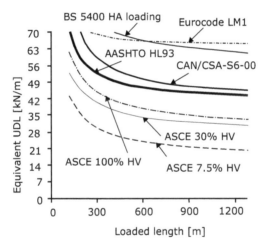

Figure 5.2 Equivalent uniformly distributed loading (UDL) for a range of long-span load models: (a) one traffic lane; (b) four traffic lanes (after Nowak et al. 2010).

focused on the three-lane lower deck of the San Francisco-Oakland Bay Bridge for two seven-hour periods in 1951. The maximum lane load at observed speeds during the two days was 2.12 kN/m (145 lb/ft). To simulate full-stop conditions, the bumper-to-bumper gap spacing between vehicles was reduced to 2.44 m (8 ft), a value determined by field measurements, which resulted in a maximum lane load of 3.6 kN/m (250 lb/ft). This pattern of mixed traffic was not thought to be suitable for a general load model due to the high ratio of cars to Heavy Goods Vehicles (HGVs) present in the traffic mix. Two further traffic mixes were therefore proposed – the first was a military convoy occupying the entire length of the structure while the second took HGV and bus traffic only with cars excluded. These two cases resulted in lane loadings of 6.6 kN/m (450 lb/ft) and 6.9 kN/m (472 lb/ft) respectively, significantly less than the 9.34 kN/m (640 lb/ft) specified by the then AASHO code (Ivy et al. 1954). The final proposed load model took account of the fact that the average weight of HGVs on the San Francisco-Oakland Bay Bridge was approximately 20% less than those in other locations due to the short haul nature of vehicles using the crossing, with 50% of the HGVs estimated to be partly loaded or empty. Loading was stepped down with increasing span length. For spans above 366 m (1200 ft) a constant load of 8.2 kN/m (560 lb/ft) was specified.

5.2.1.2 Lions Gate simulation studies

The next significant work relating to long-span bridge loading in North America arose from the need to validate the original design loading and set a weight limit for HGVs using the Lions Gate Bridge suspension bridge (Buckland et al. 1975, Navin et al. 1976). While an analytical approach was used to determine an upper bound for the critical loading, the study mainly focused on the simulation of a series of congested events. The simulation process was comprehensive and considered most of the key parameters affecting long-span LEs. It allowed the derivation of bridge-specific load levels for structures on differing routes such as a recreational route or an area with a high industrial concentration.

To conduct the simulations, vehicle data was gathered from a seven-day traffic count at selected hours in December 1974, with additional information gathered from a camera placed on one of the suspension bridge towers. Vehicle weights were inferred from the registered maximum GVW of each vehicle based on previous traffic studies, with cars and buses assumed to have constant weights and lengths. The model fed these vehicles randomly onto the bridge and simulated congestion events, such as mechanical failures, one- or two-lane accidents and head-on collisions. An average of 194 incidents a month were recorded from 1972–1974 at the bridge site, with the majority related to minor mechanical incidents and just 21 related to accidents. Speed-spacing relationships between the vehicles were determined from time-lapse photographs, while stationary full-stop conditions were simulated using a constant bumper-to-bumper gap of 1.42 m (4ft 8") derived as the average value of 100 measurements made between cars during rush hour traffic. The location of the accident, time of day, and time to clear the event were also accounted for in the simulation model. Traffic was then scanned for maximum total load, bending moment, and shear force for a range of loaded lengths. Having found the maximum loading in one lane, maxima were searched for in adjacent lanes. The maximum value during a simulated 90-day period was calculated and plotted on a Gumbel probability paper plot to predict the characteristic maximum LEs for a return period of five years (see Chapter 3).

Based on the simulation results, a load model was developed, consisting of a uniformly distributed lane load, whose intensity decreased with increasing loaded length, and a concentrated load that increased with loaded length. For the Lions Gate bridge, with a loaded length of 830 m (2725 ft) the resultant load model was approximately 50% of the AASHO load model at the time. When compared with later load models, the load levels recommended for the Lions Gate bridge are low. One reason for this is that the GVWs of the trucks used in the study were low with no major peak at heavier weights of 300 kN and above, in contrast to later data collected in the UK in 1990 and at Auxerre for development of the Eurocode in 1986. A second reason is that due to the nature of the traffic on the Lions Gate bridge, the percentage of HGVs in the measurements was low at about 3.5% of the traffic. Further studies at the Second Narrows Bridge in Vancouver found HGVs to be 7.5% of the traffic flow, and the load model was later extended to apply to general bridge structures with this as the baseline (Buckland et al. 1978, Buckland et al. 1980, ASCE 1981).

5.2.1.3 Current AASHTO load model

The current AASHTO traffic-load model, HL-93, described in Chapter 2, was updated in 1993. Many years later, Nowak et al. (2010) undertook a study to determine the suitability of this model for long-span bridges and to take account of the significant increase in HGV traffic patterns from 4% of the highway traffic population in the 1990s to 8% in 2005. Using data from 29 WIM sites, a series of traffic jams were simulated where HGVs only occupied a single lane with a constant inter-vehicle gap, axle to axle, of 7.6 m (25 ft). This layout was based on video footage where it was observed that in most traffic jam situations, HGVs tended to use a single lane with other lanes empty or occupied by cars. Equivalent uniformly distributed loads were determined for a range of spans up to 1524 m (5000 ft). As bias factors, the ratio of the mean-to-nominal load intensities, for the WIM sites did not exceed 1.25 for the heaviest 75-year combination of vehicles, the HL-93 model was determined to be suitable for long-span structures.

5.2.2 Development of United Kingdom load models

The British Standard BS 5400 was issued in the late 1970s (BSI 1978). Notable for long-span structures was the increase in the minimum value of uniformly distributed loading for normal (HA) traffic loading from 5.8 kN/m to 9 kN/m for loaded lengths above 380 m. While not providing an explicit basis for this increase, Henderson et al. (1973) note that traffic may be at closer gaps than assumed for the 'characteristic arrangement.' Dawe (2003) notes that for spans above 30 m, the vehicle mix in the assumed convoy of 235 kN HGVs was updated to intersperse lighter 95 kN and 49 kN vehicles. While heavier vehicles were permitted at the time, their LEs were not considered more onerous that the 235 kN vehicles.

A major review of long-span loading in the UK was instigated as part of the assessment and strengthening works undertaken on the Severn Suspension Bridge in the 1980s (Flint & Neill Partnership 1986). This study found that the highway traffic loading being experienced by the bridge was not adequately covered by the loading requirements of the then current code, BS 5400. Initial studies showed that spacings between vehicles were significantly less than those previously assumed. The nature of HGV traffic had also evolved with a significant increase in the numbers of HGVs with a GVW above 275 kN (28 tonnes), combined with a reduction in the number of HGVs with a GVW less than 108 kN (11 tonnes).

Following the review findings, a detailed study was undertaken to simulate traffic flows on long-span bridges. Traffic flows and mixes were based on records taken at various locations in the UK in 1980 and were intended to represent a heavily trafficked commercial route. Extrapolations were made to estimate future traffic levels for the target year of 1990. In deriving the load model, the period from 7–8 am was found to give the critical combination of vehicle flow and percentage of HGVs for long-span structures. Jams were assumed to last for one hour, with a jam frequency determined based on the vehicle flow. For the Severn Bridge case, this equated to two jams per carriageway per year. Several vehicle gap models were used, with later models using a truncated Normal distribution for loaded lengths between 100 and 1000 m to simulate observations. For shorter and longer loaded lengths, constant bumper-to-bumper gaps of 1.35 and 2.7 m were adopted. A simplified lane selection technique was used to distribute vehicles between lanes if the queue in one lane greatly exceeds another.

LEs were extrapolated to a return period of 2400 years, and an updated load model with a uniformly distributed lane load and 120 kN concentrated point load developed. The load model was significantly heavier than earlier load models, with a lane loading of 17.2 kN/m at a loaded length of 1600 m. It was introduced as BD 37/88 in 1988 and remained in place until superseded by the Eurocode in 2010.

5.2.3 Development of other European load models

5.2.3.1 Eurocode

The Eurocode load model for bridge traffic loading, EC1, is described in Chapter 2. It was developed for loaded lengths up to 200 m, but has been adopted for longer lengths (Guo & Caprani 2019). The code was developed using WIM data recorded over a two-week period on the A6 motorway near Auxerre, France, deemed to be representative of a heavily trafficked European highway. Free flowing, congested, and jammed scenarios were considered. Congested scenarios considered traffic moving at slow speeds (5–10 kph), while jammed scenarios were simulated by reducing the inter vehicle axle-to-axle gaps to 5 m. For both congested and jammed scenarios, traffic on the first lane was represented by HGVs only, i.e. all cars were removed. Lane 2 was considered to have Auxerre slow lane traffic flowing or jammed, with lanes 3–4 considered to have Auxerre slow lane traffic flowing (Bruls et al. 1996a). LEs calculated from these traffic scenarios were extrapolated to a return period of 1000 years (Chapter 3).

While early calibrations proposed that the uniformly distributed loading should reduce with increasing loaded length, the final model specified a constant intensity (Bruls et al. 1996b). As a result, this model is conservative for longer spans. The Eurocode allows for the use of modification or alpha factors on the load intensities to allow for variations between countries and regions. As a result, there are significant variations between countries, depending on their local traffic characteristics.

5.2.3.2 Storebælt East Bridge, Denmark

The base studies to determine the traffic loading for the design of the 1624 m main span Storebælt East Bridge in Denmark modelled traffic as white noise, the mean and intensity of which varied with the traffic situation (Ditlevsen & Madsen, 1994). While the bridge was designed using Eurocode standards, the magnitude of the design traffic loading was reduced from 9 to 5 kN/m^2 in the main lane. The recommended Eurocode loading of 2.5 kN/m^2 for adjacent lanes was retained, but dropped to 1 kN/m^2 for lanes travelling in the opposite direction (COWI 1990).

5.2.4 Asian long-span load models

5.2.4.1 China

For all bridges, the current Chinese standard, JTG D60-2015 stipulates the use of a truck load (for local effects) or a lane load of 10.5 kN/m. For long-span bridges, with loaded lengths over 150 m, a reduction in the lane load is permitted. For spans between 150 and 400 m, the lane load is reduced by a factor of 0.97. For each 200 m increase in loaded length, the reduction factor decreases by 0.01 until reaching 0.93 for lengths greater than 1000 m. In addition, the design load model is based on two lanes, and for single lanes a lane factor of 1.2 is applied. Conversely, for more than two lanes, the factor is less than 1.0 (Zhou et al. 2018).

5.2.4.2 Japan

Guidance on the design of long-span bridge structures (loaded length > 200 m) in Japan is provided by the Honshu-Shikoku Bridge Authority's 'Superstructure Design Standard' (HSBA 2001). First introduced in the 1970s, it has been updated several times. The current load model consists of two uniformly distributed load components, p_1 and p_2. Component p_1

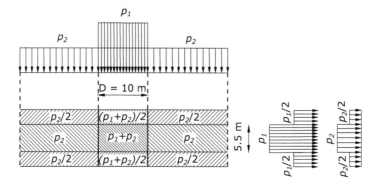

Figure 5.3 Japanese long-span bridge load model (HSBA 2001).

accounts for local effects and is applied over a length of 10 m, and has values of 12 kN/m² generally, but 10 kN/m² for bending moment on loaded lengths < 200 m. Component, p_2 is applied over the full loaded length, and has a value that reduces with loaded length, but is 3 kN/m² for lengths between 130 and 500 m. The summation and transverse values of these components is illustrated in Figure 5.3. While there is no upper limit specified, the load intensity is 2.40 kN/m² at a loaded length of 1500 m. The initial baseline for the load model is 20 kN/m, equivalent to five 200 kN HGVs on a 50 m road segment. However, based on the results of traffic studies, this value is reduced significantly in the code.

5.2.4.3 Korea

As the number of long-span bridges began to increase in Korea, initial design guidance was introduced by the Korea Society of Civil Engineers in 2006 using traffic data from static weigh stations (Hwang et al. 2013). Updates to this load model have been proposed using a more extensive WIM database, combined with a number of deterministic traffic congestion 'scenarios' (Hwang & Kim 2019). In these scenarios, the axle-to-axle spacing between vehicles is reduced to a constant value of 4.5 m, with the traffic mix consisting of HGVs only in the slow lane, or an even mix of 30% HGVs in all lanes. Simulation results are then used to calibrate the uniform load component of the existing short- to medium-span load model, which reduces in intensity as loaded length increases.

5.3 MODELING LONG-SPAN TRAFFIC LOADING

5.3.1 Traditional approaches

The fundamental work leading to current traffic-load models in codes of practice variously accounts for the modeling of:

- Vehicle characteristics;
- Vehicle fleet composition;
- Gaps between vehicles;
- Statistical extrapolation.

The models developed for these aspects of the problem are quite similar to the corresponding models for short- to medium-span bridges. However, there are some significant differences resulting from the more important role of the average vehicle stream loading (load per unit length) for longer loaded lengths.

For modeling the vehicle characteristics, the representation of truck GVW continues to be significant for long-span bridges. Truck weights and lengths are determined as discussed in Chapter 2. Although cars and other lighter vehicles do not contribute significantly in terms of weight, their presence has an important influence on the spatial layout of heavier trucks and hence on the average intensity of load. These light goods vehicles (LGVs) are represented in different models by:

- A UDL of 2.0 kN/m (Harman et al. 1984);
- A fixed length and weight of 5.5 m and 3.5 t (Flint and Neill 1986);
- A fixed length and weight of 6 m and 4 t (OBrien et al. 2010);
- Taking the LGV weights and lengths directly from the recorded traffic (Nowak et al. 2010);
- Creating a set of standard LGVs (Croce & Salvatore 1998, 2001);
- Using a statistical distribution of LGV lengths and weights (Crespo-Minguillón & Casas 1997).

In recognition of the small role of their weight, cars are disregarded completely in some models (Vrouwenvelder & Waarts 1993, Croce & Salvatore 1998, 2001). Other models represent cars by:

- A UDL of 2.0 kN/m (Harman et al. 1984);
- A fixed length and weight of:
  - 4.9 m and 1.59 t (Buckland et al. 1978, Buckland 1981);
  - 4.5 m and 1.2 t (Flint and Neill 1986);
  - 4.3 m and 0 t (i.e. for spacing purposes only) (Vrouwenvelder & Waarts 1993);
  - 4.5 m and 1.5 t (OBrien et al. 2010);
- A deterministic car length and a mean car weight (Ditlevsen & Madsen 1994);
- Using a statistical distribution of car lengths and weights (Crespo-Minguillón & Casas 1997).

Vehicle arrival is typically assumed to be independent and random. This is done for the background studies to the Eurocode (Bruls et al. 1996, Flint & Jacob 1996). Other studies model truck arrivals statistically, such as Vrouwenvelder & Waarts (1993). In a notable study Crespo-Minguillón & Casas (1997) used a Markov vehicle arrival process with the transition matrix estimated from traffic measurements. In recognition of the potential for intrinsic correlations in vehicle arrivals, some studies maintain actual recorded vehicle patterns, for example, Ivy et al. (1954) and Nowak et al. (2010).

For long-span bridge loading, the gaps between vehicles are critical. However, similar to other phenomena in the problem, the longer loaded lengths mean that the significance of individual variations is far less than for short- to medium-span bridges, but instead the average gap is critical to the average load intensity. In recognition of the congested nature of critical loading events, this gap is typically quite small and so a fixed inter-vehicle gap (back axle of front vehicle to front axle of following vehicle) is often used. Such studies include:

- Nowak et al. (2010) who use a gap of 7.6 m between trucks;
- Nowak & Hong (1991) who assume gaps of both 15 ft (4.57 m) and 30 ft (9.14 m);
- Vrouwenvelder & Waarts (1993) who use two gap models: a gap of 5.5 m for distributed lane loads and a variable gap of 4 to 10 m for full modeling, and;
- For the Eurocode background studies, where a fixed gap of 5 m is used (Bruls et al. 1996, Flint & Jacob 1996, Prat 2001).

Other studies use statistical modeling for the gaps:

- Bailey (1996) uses a beta distribution with a mode corresponding to a bumper-to-bumper gap of approximately 6.4 m, with a minimum of 1.2 m;
- Harman et al. (1984) randomly select gaps from observations, using a minimum of 1.5 m;
- Croce & Salvatore (1998, 2001) use a truncated exponential distribution with a lower bound of 5 m and upper bound of 10 m;
- Caprani (2012a) uses a Normally-distributed gap, calibrated to lane flowrates and car-truck percentage mix.

Some studies recognize that vehicle gaps and speeds are inter-related and relax the assumption of slow-moving congested traffic, using vehicle speeds to infer gaps. These include Ivy et al. (1954) and Buckland (1981), who impose a minimum gap of 1.5 m for stationary vehicles. Similarly, Vrouwenvelder & Waarts (1993) randomly vary the gaps between stationary vehicles in the range 1 to 5 m, and for moving vehicles in congested flow, vary it between 4 to 10 m based on the velocity. Further, Carey et al. (2018) acknowledge the different gaps maintained between car-to-car and truck-to-truck, as speed varies. In contrast to the preceding studies which use a fixed gap for the vehicles transversely, some studies allow gaps to vary in the transverse direction, such as Crespo-Minguillón & Casas (1997).

The models described so far have been limited in the quantity (or duration) of LE data that can be generated. As such, statistical extrapolation of these short run results is required, as it is for short- to medium-spans (Chapter 3). Nevertheless, the different nature of long-span bridge loading could mean that refinements of the extrapolation procedures are required, and so it is instructive to consider how the literature has considered this aspect so far. Approaches based on extreme value theory include Buckland (1981), Flint and Neill (1986), and OBrien et al. (2010). Interestingly, Crespo-Minguillón & Casas (1997) propose an extrapolation based on a Generalized Pareto Distribution fitted to the tail of the maximum-per-week LEs. The Normal distribution has been used in some studies – for example, Vrouwenvelder & Waarts (1993) for distributed lane loads, and Cremona & Caracilli (1998) – determine an optimal fitting of Rice's formula for the upper Normal tail. Some alternative approaches based on stochastic processes have also been developed: Croce & Salvatore (1998, 2001) determine an analytical expression for the cumulative distribution function of the maximum LE caused by vehicle convoys over a given time interval; and in Ditlevsen & Madsen (1994) the mean intensity of the white-noise traffic-load field is used to calculate the maximum LEs for a given return period.

Long-span bridge loading is complicated by a lack of suitable data. Most Weigh-in-Motion technologies, which are used extensively to find data for bridge traffic loading (Chapter 2), do not work at low speeds or stop/start situations, typical of congestion. Further, vehicle speed and inter-vehicle gaps rely on induction loops, which also do not work at low speeds (Klein et al. 2006). Consequently, there is little data on congested traffic available for study. A few studies have attempted to use cameras to capture relevant data, but the problem is complex (Micu et al. 2018, 2019, 2020, OBrien et al. 2012, 2018). In most studies to date, the WIM data used for long-span bridge loading is limited to free-flowing traffic, and so does not adequately capture vehicle gaps and any dependency between vehicle arrivals, such as may be caused by lane-changing events prior to the congested region. As a consequence of these extensive limitations, an alternative holistic methodology has emerged over the last decade or so.

5.3.2 Traffic microsimulation

As noted, when WIM data is used in a long-span bridge loading study, it is generally recorded in free-flowing traffic conditions. As congestion is known to be the critical traffic-loading case for these bridges, there is a need to 'convert' the free-flow traffic data into equivalent congested traffic. In much past work, the measured free-flow traffic data has been 'collapsed' to create the congested traffic stream. That is, the gaps between vehicles (as measured), have been reduced to minimum values. This approach has several problems and can actually be non-conservative. For example, it results in slow lane queues becoming much longer than those in the faster lanes because it neglects the lane changes that actually occur as free flowing traffic becomes congested. Worse, as it is lighter, faster vehicles that tend to change lanes, it neglects the accumulation of consecutive heavy vehicles in the slow lane, which leads to very onerous loading conditions (Enright et al. 2012). Recognizing these problems in long-span bridge loading, over the last 15 years or so, an alternative approach to the collapsing of inter-vehicle gaps has emerged: the use of traffic microsimulation (Hayrapetova 2006, Caprani & OBrien 2008, OBrien et al. 2010, Chen & Wu 2011, Caprani 2012, Enright et al. 2013, OBrien et al. 2015, Ruan et al. 2017).

Traffic microsimulation is used extensively in traffic engineering to make predictions on the impact of new roads, property developments, or altered road layouts for example. It is microscopic in scale because it models the movements of individual vehicles (rather than meso- or macroscopic modeling of traffic flow itself). As such, it is an ideal approach for understanding traffic loading on bridges; the reason for this is twofold, as it reveals: (a) the precise location of each vehicle at any point in time, and; (b) the mix between the various vehicle types, between lanes and within lanes.

There are a range of traffic microsimulation modeling types, often termed car-following models. Cellular automata approaches such as the Nagel-Schrenkenberg models are good for overall traffic flows, but represent the road as a set of cells (e.g. 7.5 m long). This limits the placement of vehicles to the spatially discrete cells, which leads to inaccuracy of LE calculation, especially for peaked influence lines, although refinements are possible (Ruan et al. 2017). There are spatially continuous models, such as the Gipps Model, Optimal Velocity Model, or Intelligent Driver Model, which can position vehicles anywhere on the influence line. Both modeling types are progressed through discrete time steps. One of the spatially continuous driving models, the Intelligent Driver Model (IDM), is introduced here in detail, though similar considerations apply for all driving models.

The IDM is a car-following model, which has a modest number of physically meaningful parameters, is collision-free, and has provided good matches with real macroscopic congested traffic (Treiber et al. 2000a, Treiber et al. 2000b, Helbing et al. 2009). It has also been calibrated with real trajectory data (Kesting & Treiber 2008, Hoogendoorn & Hoogendoorn 2010, Chen et al. 2010) and compared to other calibrated car-following models, returning results comparable to more complex models (Brockfeld et al. 2004, Punzo & Simonelli 2005). In the IDM, longitudinal movement is simulated through an acceleration function:

$$\frac{dv}{dt} = a\left[1 - \left(\frac{v(t)}{v_0}\right)^4 - \left(\frac{s^*(t)}{s(t)}\right)^2\right]$$

(Eq. 5.1)

where a is the maximum acceleration; v_0, the desired speed; $v(t)$, the current speed; $s(t)$, the current gap to the vehicle in front; and $s^*(t)$, the minimum desired gap, given by:

$$s^*(t) = s_0 + \max\left\{Tv(t) + \frac{v(t)\Delta v(t)}{2\sqrt{ab}};\quad 0\right\}\qquad\text{(Eq. 5.2)}$$

in which, s_0 is the minimum bumper-to-bumper distance; T, the safe time headway; $\Delta v(t)$, the speed difference between the current vehicle and the vehicle in front; and b, the comfortable deceleration. Note that, when the front vehicle is faster, the desired minimum gap, s^* in Equation (5.2) can turn negative, generating an inconsistent driver behavior. Therefore, the desired minimum gap s^* is limited to the minimum bumper-to-bumper distance, s_0 (Caprani et al. 2016). There are five parameters in this model to capture driver behavior, (v_0, s_0, T, a, b), which are relatively easy to measure. For simulation purposes, the length of the vehicle must also be known.

For multi-lane highways, lane-changing is a critical issue, and it can be modelled using the MOBIL lane changing model proposed by Kesting et al. (2007). MOBIL (Minimizing Overall Braking Induced by Lane changes) can be adapted for either symmetric or asymmetric passing rules, such as those in the United States or the European Union respectively (Kesting et al. 2007). (An asymmetric passing rule means that passing is only allowed on one side (the faster lane) whereas a symmetric passing rule allows passing in the lanes on either side.) The topology of a lane change event is illustrated in Figure 5.4 where the subscript c (on the vehicle acceleration, a_c) refers to the lane-changing vehicle, o refers to the old follower (in the current lane), and n to the potential new follower (in the target lane). The tilde on the acceleration of vehicle c identifies the potential acceleration after the lane change. The front vehicles play a passive role, representing constraints on the lane change decision and affecting the acceleration of the lane-changing vehicle. All the accelerations, current and potential, are calculated according to the car-following model described by Equations (5.1) and (5.2).

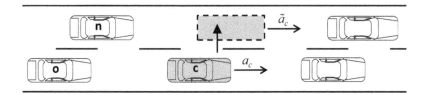

Figure 5.4 Vehicles involved in lane-changing maneuver (adapted from Kesting et al. 2007).

A lane change occurs if the incentive and the safety criteria are both satisfied. The incentive criterion is a test of the driver's desire to change lane and is represented by:

$$\tilde{a}_c(t) - a_c(t) = \Delta a_{th} - \Delta a_{bias} + p\left[\left(a_n(t) - \tilde{a}_n(t)\right) + \left(a_o(t) - \tilde{a}_o(t)\right)\right] \qquad \text{(Eq. 5.3)}$$

This means that the acceleration advantage to be gained by lane-changing, $\tilde{a}_c - a_c$, must be greater than the sum of the acceleration threshold, Δa_{th}, which prevents overtaking with a marginal advantage, the bias acceleration, Δa_{bias}, which acts as an incentive to keep to the slow lane for highways with asymmetric passing rules, and the imposed disadvantage to followers in the target lane $a_n - \tilde{a}_n$ and in the old lane, $a_o - \tilde{a}_o$, weighted through a politeness factor, p. Consequently, driver aggressiveness can be adjusted with the politeness factor. While $a_n - \tilde{a}_n$ is usually positive, representing a real disadvantage for the new follower, $a_o - \tilde{a}_o$ is usually negative, representing an advantage for the old vehicle. Thus, it is taken into account that a faster follower can pressurize its leader. Finally, the safety criterion limits the imposed deceleration to the follower in the target lane:

$$\tilde{a}_n(t) \geq b_{safe} \qquad \text{(Eq. 5.4)}$$

It is quite rare that the safety criterion of Equation (5.4) applies, as long as the politeness factor p is not too close to zero. Further details about the IDM and MOBIL models can be found in Kesting et al. (2007). Typically, the equations are discretized into 250 ms steps and the integrations carried out using an Euler scheme, but other schemes are possible (Treiber & Kanagaraj 2015).

Traffic microsimulation models must of course be calibrated to ensure consistency with the physical phenomena. Each model has distinct processes to be followed for calibration. However, some models contain parameters that are purely notional and not physically measurable (e.g. cell size for cellular automata models). A particular attraction of the IDM is that its parameters are both physically meaningful and measurable. Indeed, many studies use IDM to simulate traffic and demonstrate that the results can match observed traffic quite well (Treiber et al. 2000a, Kesting & Treiber 2008, Helbing et al. 2009, Chen et al. 2010, Caprani 2012).

A fairly representative set of parameters shown to replicate real traffic behavior quite well is given in Table 5.1 (Treiber et al. 2000a, Kesting & Treiber 2008, Caprani et al. 2012, OBrien et al. 2015). Kesting & Treiber (2008) describe the calibration of multiple car-following models (for longitudinal motion). Lipari (2013) describes the calibration process of the MOBIL lane-changing in detail. Three major constraints are considered to find the suitable set of MOBIL parameters: maintaining the ability of the IDM to reproduce congested traffic states (explained in Section 5.3.3);

Table 5.1 Example Set of Parameters for the Intelligent
Driver Model (Caprani et al. 2016)

Parameter	Car	Truck
Safe time headway, T (s)	1.6	1.6
Maximum acceleration, a (m/s²)	0.73	0.73
Comfortable deceleration, b (m/s²)	1.67	1.67
Minimum jam distance, s_0 (m)	2	2
Desired velocity, v_0 (km/h)	120±20%	80±20%
Lane change politeness factor, p	0.2	0.2
Outside lane bias factor, Δa_{bias}	0	0
Lane change threshold, Δa_{th} (m/s²)	0.4	0.4
Maximum safe deceleration, b_{safe} (m/s²)	4	4

matching the observed lane change rates for 2-lane motorways (Sparmann 1979, Yousif & Hunt 1995); and being in general agreement with the observed percentages of trucks in the fast lane (Ricketts & Page 1997, Hayrapetova 2006, OBrien & Enright 2011, Fwa & Li 1995).

Calibration of traffic microsimulation is typically done to ensure an accurate portrayal of vehicle movements. However, for long-span bridge loading, it is the LE that is of interest, and so the influence of the microsimulation model calibration parameters on loading is of interest. Lipari (2013) investigated the sensitivity of long-span bridge traffic LEs to the traffic microsimulation parameters. He found that the traffic LE is not very sensitive to these parameters, especially for lane changing. Lipari (2013) explores a range of parameter sets and flow rates, finding a variation in resulting load effects of about ±5% for quite different driving model parameter sets. This figure may provide useful guidance on the likely model error attributable to LEs determined using traffic microsimulation when used as part of a reliability analysis.

For single-lane traffic, Table 5.2 gives examples of the magnitudes of loading that traffic microsimulation indicates for different forms of congestion as derived from the results of OBrien et al. (2015). It should be noted that it is a simplified study and these values are indicative only of traffic composed of cars and 5-axle semi-trailer trucks. Nevertheless, the results confirm that for longer loaded lengths, the Equivalent UDL reduces. However, it is noteworthy that there is a significant difference in loading resulting from both different forms of congestion and the proportion of trucks. Both of these are strongly site-specific in terms of traffic composition and the relative frequencies of occurrence of the different forms of congestion. As such, the results point to the value of this form of analysis for the assessment of existing bridges, where these inputs can be measured, and a more rational, less onerous load model derived.

Table 5.2 1000-Year Characteristic Equivalent UDLs (kN/m) of Single-Lane Traffic Derived from the Total Load on the Loaded Length

	200 m		1000 m	
Type of Congestion	*20% Trucks*	*50% Trucks*	*20% Trucks*	*50% Trucks*
SGW	19.8	22.7	8.7	11.9
OCT	20.7	23.6	10.9	15.1
HCT/OCT	20.9	23.8	12.8	17.8
HCT (1)	24.8	27.0	14.8	20.3
HCT (2)	26.0	27.8	16.1	22.2
FS	22.8	31.3	18.2	26.3

SGW = stop-and-go waves, OCT = oscillating congested traffic, HCT = homogenous congested traffic, FS = full stop. HCT (1) and (2) have average speeds of about 9 km/h and 5 km/h respectively. (Based on Results in OBrien et al. 2015.)

5.3.3 Lane changing and types of congested traffic

Much of the bridge engineering literature on traffic loading considers full-stop (congested) traffic to be the critical case of traffic loading for long-span bridges. For example, Ricketts & Page (1997) state that only standstill traffic is critical for bridge loading, and that this happens for only 2% of the congested time. Buckland et al. (1980) assume 800 standstill traffic events per year for the background work to the ASCE long-span bridge load model (ASCE 1981). For the background studies to the Eurocode load model, the Flint and Neill Partnership (1986) assumed one jam with a queue of vehicles at minimum bumper-to-bumper distances for every 80,000 km traveled. The rationale behind the assumption of full-stop traffic being critical is that the vehicles are at a minimum spacing. In contrast to these works, real-world observations show that congestion can take different forms. Indeed, in the traffic engineering community, traffic conditions are not taken as a binary choice between free-flow traffic and congested-flow traffic. Rather, traffic states occur on a spectrum, with many forms of congestion. Key states of interest for long-span bridge loading are:

- Triggered stop-and-go waves: individual vehicles oscillate between low speeds (~5 km/h) and speeds of about 60–70 km/h, over wave-lengths of about 600–1000 m;
- Oscillating congested traffic: vehicle speeds oscillate but do not exceed 30–40 km/h over wavelengths of about 200 m;
- Homogenous congested traffic (HCT): constant slow-moving jam of about 5–10 km/h;
- Full-stop conditions.

The occurrence of these traffic types is mostly governed by the difference between traffic flow demand and road capacity, termed the bottleneck

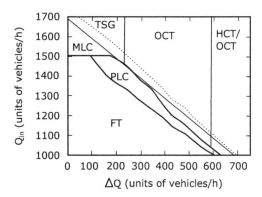

Figure 5.5 Phase diagram, indicating regions of different traffic types (adapted from that of Treiber et al. 2000). Q_{in} = inflow, ΔQ = bottleneck strength, TSG = triggered stop-and-go, OCT = oscillating congested traffic, HCT = homogenous congested traffic, MLC = moving localized clusters, PLC = pinned localized clusters, FT = free-flow traffic.

strength. Usually, the higher the bottleneck strength, the lower the road capacity at that location. Treiber et al. (2000a) discuss the relationship between bottleneck strength and different congestion types using a phase diagram; see Figure 5.5.

The bottleneck strength (or the dynamic capacity difference), ΔQ, is the local decrease in the traffic capacity:

$$\Delta Q = Q_{out} - Q'_{out} \qquad \text{(Eq. 5.5)}$$

where Q_{out} is the maximum dynamic road capacity on the (hypothetically) free road and Q'_{out} is the maximum dynamic road capacity with the bottleneck in place. It is interesting to note that the road capacity (vehicles per hour) is a function of the traffic composition since larger vehicles take up more road space (OBrien et al. 2015). Bottlenecks are caused by road features and events such as speed limits, steep hills, accidents, heavy rain, driver distractions, and so on. To simulate congestion, bottleneck strength can be introduced into an IDM at a specific road location by either directly decreasing the desired speed v_0 (e.g. a speed limit) or by increasing the safe time headway T (driver response time, leading to bigger desired gaps). Both will reduce the capacity of the road, but Treiber et al. (2000a) suggest that the latter is more effective.

Lane changing is a key component of traffic microsimulation for bridge traffic loading. In the past, it was common practice in the long-span bridge traffic-loading literature to:

- Assume vehicle arrivals are independent, based on traffic composition (e.g. Flint & Jacob 1996, Crespo- Minguillón & Casas 1997), or;

- Collapse free-flowing traffic streams into closely-spaced trains of vehicles (e.g. Nowak et al. 2010, Vrouwenvelder & Waarts 1993).

However, as noted earlier, these assumptions neglect the impact of lane-changing on the resulting traffic LE. Due to the differing vehicle and driver parameters, cars tend to pull out from behind slower moving trucks to over-take. Lane changing increases significantly close to the traffic breakdown front (in the transition from free-flowing to congested), as illustrated in Figure 5.6. This filtering of cars from between trucks leads to the forma-tion of long platoons of trucks, which can be significant for bridge traffic loading. While the effect depends on the percentage of trucks in the traffic stream, the LEs can increase by up to 25% once lane changing is allowed for (Caprani et al. 2016).

The influence of lane-changing and other factors particular to multi-lane traffic on the resulting load effects are highlighted in Table 5.3, derived from the results in the study by Caprani et al. (2016). The traffic used in this study is simplified, as per the single-lane analysis in Table 5.2. Again, con-sistent with previous research, the equivalent UDL for two-lanes is found

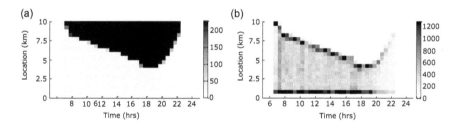

Figure 5.6 Distribution of lane change frequency (events/km/hour) from traffic micro-simulation of a heavily trafficked site. (a) Vehicle density (veh/km); (b) lane change rates (events/km/hour). Low levels of lane-changing indicate highly congested traffic (lightly-colored region). High levels of lane changing can be seen in advance of the congested traffic zone (dark banded region) (after Carey et al. 2012).

Table 5.3 1000-year Characteristic Equivalent UDLs (kN/m) of Two-Lane Traffic Derived from the Total Load on the Loaded Length

	200 m		1000 m	
Inflow (Veh/Hour)	HCT (2)	FS	HCT (2)	FS
1250	35.0	39.8	19.3	33.3
2000	41.2	38.5	28.1	32.7
3000	47.5	45.1	27.7	32.9

HCT = homogenous congested traffic, FS = full stop, HCT (2) has an average speed of about 5 km/h. (Based on Results in Caprani et al. 2016.)

to be less than twice that for a single lane (e.g. compare HCT(2) for 200 m span of 47.5 kN/m to 27.8 kN/m in Table 5.2), and this phenomenon is consistent with the multi-lane reduction factors found in many codes. However, the benefit here is that the site-specific characteristics can be captured for improved loading estimates. It is also interesting to note that once beyond the road capacity of about 2000 veh/hr, congestion occurs, and there is typically little influence of additional vehicles entering the road. Finally, it is interesting to compare the extreme loadings resulting from the slowly moving HCT traffic, with the single observational Full-Stop traffic.

5.3.4 Extreme traffic load effects

Caprani et al. (2012) and OBrien et al. (2015) study the influence of the different types of congestion on long-span bridge traffic LEs. The findings counter some of the past assumptions made in traffic-loading research. While it is confirmed that the full-stop traffic state, with its most closely spaced vehicles, is the critical situation *on average*, it is found that the HCT state is more critical for bridge loading, considering the number of occurrences. That is, for each hour of exposure (event occurrence) on a 1 km section of road, a full-stop loading event will have only one set of vehicle weights. Homogenous congested traffic, on the other hand, traveling at, say, 10 km/h, will subject the bridge to ten independent combinations of vehicle weights in that hour. Given the random nature of vehicle weights, this increases the probability of at least one of these combinations being more extreme than the single set of weights present in the full-stop state. Consequently, considering both the frequency of occurrence of these loading events, and the statistical extrapolation to a suitable return period, it is necessary to consider more than just one type of congestion (OBrien et al. 2015, Caprani et al. 2016).

A detailed formulation considering the extrapolation of multiple forms of traffic state, along with their relative frequencies of occurrence for a specific site, is given by Lipari et al. (2017). The problem is complicated by the need for distributions of load effect (z) for a given type of congestion. And because the type of congestion depends on the severity of the incident (i) that occurred (to trigger congestion) and the traffic composition and flow rate (q) at the time, a statistical distribution emerges, $f(z,i,q)$, from which the overall distribution of load effect can be determined by integrating over the frequencies of incident severity and flows:

$$F(z) = \iint f(z,i,q) f_{i,q}(i,q) \, \mathrm{d}i \, \mathrm{d}q \qquad \text{(Eq. 5.6)}$$

This formulation requires extensive scenarios of traffic simulation to be conducted for the site, considering its vehicle mixes and flows, and is best suited to advanced assessments of existing long-span bridges or the development of new load models.

The critical form of traffic loading for a long-span bridge LE is also sensitive to the shape of the influence line for the LE under consideration. Section 5.4 illustrates the influence lines for several forms of structure and LE. Clearly both adverse and beneficial regions of loading exist. It is common and conservative practice to ignore the contribution of loading to any beneficial part of the influence line, and this is included in many codes of practice (e.g. AASHTO, Eurocode, Australian Standard). Statistically, the problem with this approach is that it is a joint extreme: it combines both extreme loading values and an extreme loading topology, without regard for the joint probability of occurrence. While it is conservative, this approach is having a significant impact on the design and assessment of long-span bridges (Guo & Caprani 2019).

Considering the types of congestion, neither full-stop nor HCT congestion states, which are being represented by the notional models in codes of practice, have innate patterning. In contrast, triggered stop-and-go waves have both dense and free-flowing zones, and of course these can become coincident with adverse and beneficial parts of an influence line, leading to a critical loading situation. Thus, while it is unduly conservative to completely neglect loading on the beneficial parts of the influence line, it is also not reflective of the critical traffic phenomenon to apply the full-stop or HCT traffic-load model only to the beneficial parts of the influence line. Based on Turkstra's Rule for load combination, and extensive simulations of traffic scenarios and influence lines, Guo & Caprani (2019) propose that while the adverse parts of the influence line are loaded with the lifetime extreme traffic loading, the beneficial parts should be loaded with a daily mean traffic loading. This daily mean traffic reflects free flowing conditions and can be determined from the load model following some statistical considerations. As a rule of thumb, this 'beneficial loading' is approximately 10% of the extreme loading.

5.3.5 Conclusions and future directions

At the time of writing, significant challenges remain in the modelling of load on long-span bridges. Assumptions about gaps between vehicles in congested traffic and full-stop states are critically important but there is a dearth of gap data in the literature. The mix of vehicle types, both within lanes and between lanes, is also important. While there are now vast quantities of WIM data, they are only available for free-flowing conditions. Microsimulation addresses this issue but there is a strong need for validation studies to confirm the results of simulations. Recent work has attempted to provide these validations using new measurement approaches.

5.3.5.1 Inter-vehicle gap data and mix of vehicle types

Micu et al. (2018, 2019) use a camera mounted on the tower of a suspension bridge, to collect data on inter-vehicle gaps and the mix of vehicle types in two lanes, over a five-month period. Image analysis software was used to identify vehicles in the images and to estimate their lengths; see Figure 5.7. A correction was applied to allow for perspective and the inclusion of the vehicle fronts in the images. One year of data from a WIM system located nearby was used to establish the relationship between vehicle length and weight. They extracted histograms of vehicle weights for each vehicle length. These are used, in a Monte Carlo simulation process, to generate 'typical' weights for the vehicles in the images. In a numerical study, Micu et al. (2018) show that the inference of weight from length, while inaccurate for individual vehicles, has little effect on the accuracy of the characteristic maximum LEs. This approach overcomes the challenge of measuring inter-vehicle gaps and finding the mix of adjacent vehicles in multiple lanes. However, in the five months of measurement for this study, there were only two fully jammed loading events, so further measurement campaigns are needed to comprehensively examine inter-vehicle gaps in full-stop conditions.

5.3.5.2 Load effect calculation

For long-span bridges, the calculation of LEs should consider the evolution of vehicle gaps and the changing mix of vehicle types as the traffic stream crosses the bridge. As noted in Chapter 3, many studies have used a time-based calculation in which vehicles maintain their velocity measured at a

Figure 5.7 Image detection of vehicles: a) example 1; b) example 2; c) example 3 (after Micu et al. 2019).

point. This approach is relatively straightforward to implement, but it can lead to physically impossible overlapping of vehicles further down the road due to differing vehicle velocities. It can be solved by imposing a constant speed on all vehicles, thus freezing the time headways. However, the selection of this constant velocity is difficult, especially, for situations where the speeds are changing significantly (e.g. stop-and-go waves). A space-based calculation in which the traffic topology is extracted at each point in time, and the resulting LEs calculated, is clearly preferable. However, this approach is difficult to implement as it requires the merging of both traffic microsimulation and bridge LE calculation algorithms. Nevertheless, it facilitates the precise determination of LEs, and allows for all forms of traffic evolution on the road. Although clearly precise, it is not necessary to consider every axle on the bridge surface: the approach can be made more efficient by considering the curvature of the influence lines, and representing the loading in alternative ways (e.g. point loads, uniformly distributed loads, etc.) that does not compromise accuracy, as is proposed by Zhou et al. (2018).

5.3.5.3 Influence of truck percentage

Patterns of truck travel are often significantly different from cars, particularly for urban bridges where the flow of cars tends to be strongly influenced by commuters. Carey et al. (2018) and Micu et al. (2020) identify different regimes of traffic, of which two are significant for bridge loading:

1. While total flow is low during the night to early morning the percentage of trucks can be very high: Carey et al. (2018) find truck percentages of up to 75% on an inter-urban route near Wroclaw, Poland, while Micu et al. (2020) find 25% for an urban bridge near Edinburgh, Scotland. In these situations, the occurrence of a bottleneck due to an incident or lane closure due to road works can then cause a very severe loading situation. However, the probability of this event can be considered as quite low.
2. During the day, truck percentages can fall substantially, as flow rates increase significantly. This was observed in both Carey et al. (2018) and Micu et al. (2020). In such scenarios, the lower truck percentages reduce the intensity of an extreme loading event but the probability of such an event is considerably higher.

Both these scenarios (and other potential intermediate scenarios) should be considered in the calculation of extreme loading on a long-span bridge, likely following a total probability approach similar to Equation (5.6).

5.3.5.4 Future developments

Road transport is rapidly evolving, and the last few decades have seen a large increase in the use of Intelligent Transportation Systems. More recently, automated vehicles are appearing on roads, and it seems only a matter of time until Connected Autonomous Truck Platoons are traversing highways. The benefits to industry are reduced labor costs and fuel consumption, and improved safety. However, such automated truck platoons are likely to have an onerous effect on existing bridge infrastructure due to their small inter-vehicle gaps at full highway speed. Effectively it seems that such developments will entail congested traffic-type loading but with free-flow traffic dynamic effects. This is a critical combination not yet considered in codes of practice or Autonomous Truck Platoon trial guidelines (Caprani 2018). Some initial research is being done on the impact for highway bridges (e.g. Yarnold & Weidner 2019, Sayed et al. 2020), but it is clear that there remains a significant knowledge gap.

Fortunately, the same technology enabling closely spaced vehicles can be used to mitigate their effects on critical bridges. Caprani (2012a) suggests the use of Bridge-to-Vehicle communication to reduce the LE that a bridge experiences under congested traffic by adjusting truck driving parameters. Lipari et al. (2017) take this further, developing a detailed gap control method for the IDM to mitigate bridge traffic loading. They find that the concept is highly effective and can reduce loading up to 45% for 50 m control gaps and 90% truck compliance.

5.4 CASE STUDIES

5.4.1 Model and traffic basis

The cellular automata microscopic traffic simulation developed by Ruan et al. (2017) is used to study traffic LEs on several long-span bridges, including a regular two-pylon cable-stayed bridge, a six-pylon cable-stayed bridge, and a three-pylon suspension bridge. The model calibration and validation of the simulation technology are based on WIM traffic data collected from a highway in China. Two-lane unidirectional traffic, and seven consecutive days of WIM data measured in 2009 are utilized. The data is assessed for accuracy and filtered using an approach typical of those explained in Chapter 2 (Ruan et al. 2017). After filtering, the daily traffic volume is approximately 14,000–16,000, which is typical of two-lane traffic in China. Based on the analysis of the data, the given parameters of the microsimulation model are shown in Table 5.4. This parameter set governs the longitudinal and lane-changing movement of vehicles across the bridge, i.e. the car-following and lane-changing behaviors.

5.4.2 Two-pylon cable-stayed bridge

A two-pylon cable-stayed bridge is first considered, as depicted in Figure 5.8. It uses a flat separated prestressed concrete box girder carrying six bidirectional lanes with a total width of 27.42 m span. Bending moments at three locations are investigated: in the deck of the third support pier from the

Table 5.4 Given and Calibrated Parameters in the Microsimulation Model (Adapted from Ruan et al. 2017)

	Parameters	Value
Given	Lane number	2
	Cell size	5 m
	Time step	0.1 s
	Unit deceleration	$[0-1]\,\text{m/s}^2$
Calibrated	Random deceleration factor	0.01
	Lane-changing probability from slow to fast lane	0.01
	Lane-changing probability from fast to slow lane	0.01

right (LE1), at mid-span of the main span (LE2), and at the base of the right pylon (LE3). Using the WIM data and traffic microsimulation, time histories of those LEs are obtained. As the proportion of heavy vehicles is of concern, combinations of different AADT (Annual Average Daily Traffic volume), with various percentages of heavy vehicles, are investigated. The LEs obtained are compared with design values calculated using the traffic-load model in the current design code for bridges in China (JTG D60-2015).

As shown in Figure 5.9(a), the LEs, which are presented in terms of 98% quantiles of the ratios with respect to the design value, increase with the proportion of heavy trucks in the traffic. However, when the Average Daily Truck Traffic (ADTT) is increased to 80,000 per day (Figure 5.9(b)), the curves tend to 'flatten,' i.e. there is a reduced sensitivity to the percentage of heavy vehicles. There is also a greater variability in the extreme LEs when the percentage of heavy trucks is greater: for 20% heavy trucks, the LEs are between about 0.2 and 0.3, but when this increases to 90% heavy trucks, they are in the range from about 0.35 to 0.6. It should also be noted that the sensitivities of LEs to the proportion of heavy vehicles vary significantly by LE.

The influence of the proportion of heavy vehicles on the characteristic maximum LEs is also investigated. The results are presented in Table 5.5 for a 5% probability of exceedance in the design life of 100 years, i.e. to a return period of about (100/0.05 =) 2000 years. As expected, the ratio increases with the proportion of heavy trucks, and has the same trend as the 0.98 quantile values discussed above. For instance, the characteristic maximum varies between 0.45 and 0.75 for 20% heavy vehicles, but the range increases to between 0.75 and 1.25 for 90% heavy trucks. The traffic volume also has a significant influence on the extreme values. For 90% heavy trucks with ADTT of 100,000, the LEs vary from 0.87 to 1.07.

5.4.3 Bridges with multiple pylons

5.4.3.1 Cable-stayed bridge

A six-pylon cable-stayed bridge with five interior spans of lengths about 428 m is considered; see Figure 5.10. The bridge is 55.6 m wide, carrying

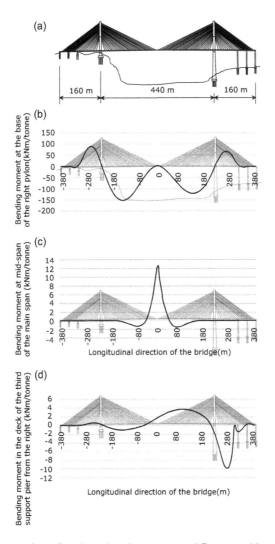

Figure 5.8 Influence lines for three bending moment LEs in a cable-stayed bridge: (a) general arrangement; (b) main deck at the top of the first pylon; (c) main deck at mid-span; (d) base of the right-hand pylon.

eight lanes of bidirectional traffic on a split-type double-box steel box girder. The pylons are single concrete columns, and the stay-cables are in four spatial planes. According to the criteria proposed by Ruan et al. (2017a), LEs are classified as global effects and partial effects based on their effective influence ranges (influence ordinates no less than 20% of the maximal absolute influence value) being larger or less than 10% of the total length. Global LEs are further divided into sensitive effects, less sensitive effects, and insensitive effects, depending on their sensitivity to unbalanced traffic loading (i.e. onerous traffic loading on the positive ranges of the influence

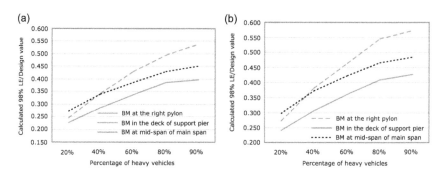

Figure 5.9 The 98% quantiles of the three LEs for various percentages of heavy vehicles: (a) for ADTT of 60000 veh/day; (b) for ADTT of 80,000 veh/day.

Table 5.5 Ratios of 2000-Year Return Levels with Respect to the Design Value for the Three LEs under Combinations of the Traffic Volumes and the Percentages of Heavy Vehicle (kNm)

ADTT (Veh/Day)	Load Effect	Percentages of Heavy Vehicle				
		20%	40%	60%	80%	90%
60,000	Bending moment in the main girder at the top of pier	0.54	0.61	0.71	0.80	0.82
	Bending moment in the main girder at the mid-span	0.67	0.80	0.85	0.90	0.93
	Bending moment at the bottom of the tower	0.58	0.72	0.88	0.95	0.98
80,000	Bending moment in the main girder at the top of pier	0.60	0.68	0.73	0.82	0.87
	Bending moment in the main girder at the mid-span	0.72	0.87	0.88	0.94	1.02
	Bending moment at the bottom of the tower	0.61	0.79	0.90	1.20	1.07

line but alleviative loadings on the negative ranges). Pylon bending moment is an example of a sensitive effect, characterized by substantial negative and positive zones in the influence line.

Representative global effects are considered: (i) axial force and (ii) bending moment at the base of the middle pylon. Based on the measured WIM data, eight-lane traffic was modeled according to the prepared microsimulation parameters in Table 5.4. A total of one-year simulated traffic was generated, and three traffic-loading scenarios were considered: free flowing, congestion, and the most adverse state. In modeling traffic congestion, local lane closure at the pylon-girder connection was performed, and the number of closed lanes and duration of the lane closure was based on traffic jam data collected on Chinese highways (Ruan et al. 2012). For the most

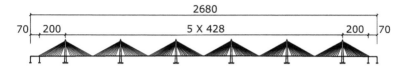

Figure 5.10 Elevation of a six-pylon cable-stayed bridge (lengths in m).

adverse state, congested traffic loads were only applied to the parts of the influence line, where it had an adverse effect on the LE. To compare with the results of the traffic-load model for design in China (JTG D60-2015), the LEs were all extrapolated to the characteristic value with a probability of failure of 5% in 100-year design life.

Two traffic parameters of Average Annual Daily Traffic (AADT) and heavy vehicle percentage (HV) are considered, where HV is defined as the percentage of vehicles with GVW in excess of 20 tonnes. From an analysis of the relevant WIM database, the AADT for the eight traffic lanes was found to approximate 38,000 veh/day, with HV approaching 12% (Ruan & Zhou 2014). The traffic volume in the highway is far from saturation, and the heavy vehicle percentage is typical for China (Zhou et al. 2020, Han et al. 2018). Therefore, sensitivity to higher AADT levels from 40,000 to 120,000 veh/day, and to HV from 10 and 20% is considered. The influence of these parameters on the characteristic LEs expressed as a ratio with respect to the Chinese design load model values is shown in Figure 5.11. As would be expected, increases in both AADT and HV cause increases in LE. The pylon moment LE of Figure 5.11(b) is highly sensitive to adverse loading, which results in this traffic state being, by far, the most critical. In comparison, LEs in Figure 5.11(a) are insensitive to adverse loading, and the three traffic states give similar results. The results indicate that traffic states significantly affect extreme values. However, even in the most unfavorable extreme state, the actual responses are still much lower than the code values, suggesting that the current Chinese design code is conservative for long-span multi-pylon cable-supported bridges. This highlights the need to establish diverse site-specific traffic-load models for the precise estimation of these structural load effects, especially for the assessment of such existing structures.

5.4.3.2 Suspension bridge

Loading is also considered on the Taizhou Yangtze River Bridge, which is a three-pylon suspension bridge with a span arrangement of 390 m-1080 m-1080m-390m, as illustrated in Figure 5.12. The main deck is a steel box section of 36.7 m width, carrying six lanes of traffic. The central steel pylon is inverted-Y-shaped (see Figure 5.12), offering some rotational resistance to the unbalanced load. Anti-sliding safety of the middle saddle is a key issue

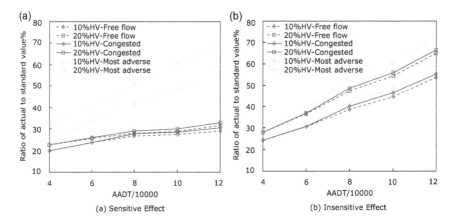

Figure 5.11 Load effects under three traffic states on a six-pylon cable-stayed bridge: (a) axial force and (b) bending moment at the base of the middle pylon (adapted from Ruan et al. 2017a).

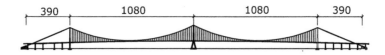

Figure 5.12 Elevation of the Taizhou Yangtze River Bridge (dimensions in m).

for the design of three-pylon suspension bridges, and traffic load is the main adverse effect, especially for an unbalanced traffic-load arrangement.

The middle saddle's anti-sliding safety is ensured when the static friction force at the saddle is in equilibrium with the difference between the two cable forces. This friction force ensures no relative movement between the saddle and the main cable. Thus, the balanced static friction coefficient due to unbalanced loading, i.e. onerous traffic loadings on one span while alleviating loadings on the adjacent one, cannot be greater than the maximum static friction coefficient between the saddle and main cable. Based on this, the limit state equation for anti-sliding behavior of the middle saddle is as follows

$$Z(R,S) = R - S = \mu - \mu_c \qquad \text{(Eq. 5.7)}$$

where Z is the safety function such that when Z is smaller than zero, the structure fails; R is the structure resistance; and S is the load response. The maximum static friction coefficient μ and balanced static friction coefficient μ_c correspond to the generic resistance (R) and loading variables (S) in the reliability assessment of anti-sliding safety for a saddle. Therefore, the anti-sliding safety of the saddle can be assessed by calculating

$p_f = P(Z \le 0) = P(\mu \le \mu_c)$. Aging over the bridge service life will have some effect on limiting the anti-sliding capacity between the saddle and main cable. Therefore, considering the material, environment, and computational model, the structural resistance probability model is built as: $R = k_m k_b k_p R_k$ with R_K for the maximum static friction coefficient, and k_m, k_b, k_{pm} for uncertainties related to the material property, the service environment, and the computational model, respectively. The statistics for these parameters can be found in Ruan et al. (2016). The main task is, therefore, to find the probabilistic model of the load effect, S, in the above equation to estimate the probability of failure.

According to a prediction of the Taizhou Yangtze River Bridge's traffic flow, the actual traffic volume is expected to be around 80,000 vehicles per day until the year 2030. Therefore, the load response for traffic volumes from 20,000 to 100,000 vehicles per day is calculated. Further information about the WIM data of the 6-lane bidirectional traffic is given by Ruan et al. (2016). Simulated traffic flows of one year with various traffic volumes is applied to the influence surface to calculate the balanced static friction coefficients (Ruan et al. 2016). The data is presented on a Gumbel paper plot (see Chapter 3) to investigate the tail tendency of the extreme load responses, as shown in Figure 5.13, in which several empirical distributions are applied to determine the best fit.

With the models for R and S, the probability of failure, $P\left[\mu \le \ln(T_1/T_2)/\theta\right]$ can be calculated by a numerical approach, such as the first-order reliability method. Table 5.6 shows the anti-sliding annual failure probability P_f

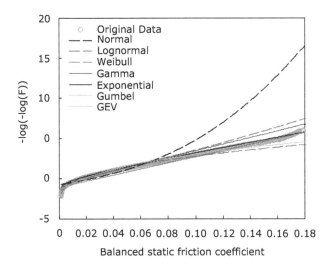

Figure 5.13 Tail fitting of balance static friction coefficient with simulated traffic volume of 80,000 veh/day (from Ruan et al. 2016a, with permission from ASCE).

Table 5.6 Reliability-Based Safety Assessment (Adapted from Ruan et al. 2016)

Traffic Volume (Veh/Day)	Resistance Normal Distribution	Load Response Exponential Distribution	P_f	β
20,000	μ = 0.5280	λ = 170.8	$< 1 \times 10^{-15}$	> 8.00
40,000	σ = 0.0317	λ = 66.4	4.66×10^{-15}	7.75
60,000		λ = 45.8	7.97×10^{-11}	6.40
80,000		λ = 33.0	4.24×10^{-08}	5.36
100,000		λ = 29.1	3.03×10^{-07}	4.99

and corresponding reliability index β for the middle saddle of the Taizhou Yangtze Bridge, and the distributions of resistance and load responses for different traffic volumes are also included. The probability of failure increases with the increase of traffic volume as expected. The increase in traffic volume is likely to lead to congestion due to many vehicles on the bridge, which may, in turn, lead to increased accidents and incidents so that the unbalanced load increases. Overall, the results show that the anti-sliding failure probability is less than the target failure probability (10^{-6} in this case) for the traffic volumes considered. It was therefore concluded that the design of the Taizhou Yangtze River Bridge satisfies the anti-sliding safety requirements for traffic volumes between 20,000 and 100,000 vehicles per day (Ruan et al. 2016).

5.5 CONCLUSIONS

Long-span bridges are amongst humanity's greatest structural achievements, linking communities over vast obstacles. While dead load usually governs the design as a whole, the discussions and case studies here show that the traffic live load is a critical consideration both in the design of new long-spans, and in the assessment of existing long-span bridges. The historical and somewhat empirical approaches to traffic-load modeling for these bridges is being replaced with the more rational and holistic modeling framework of traffic microsimulation over the last decade or so. This approach is revealing new understanding in the critical types of traffic, the importance of traffic mix or composition, and the role of load patterning on global influence lines.

Caprani & de Maria (2020) present an open global database of over 750 long-span bridges, and provide an analysis of the data. Roughly speaking, there are two clear populations of long-span bridges: those built between about 1930–1990 mainly in the United States and Europe, and those built since 1990, mainly in Asia. Those older bridges remain critical to the communities they serve, and have, in the main, required rehabilitation prior to the intended design life. More accurate traffic-load modelling for these bridges will allow precise estimates of actual safety levels and give vital information for optimizing rehabilitation measures. On the other hand, the large number of newer and even longer spans indicate that the need for new long-span bridges is unabated. Improved traffic-load models for new super-long-span bridges will help avoid unnecessary design complications, such as overly conservative load patterning, and could enable bridges where otherwise it would not be feasible. So, while there remain challenges, it can be expected that the deeper understanding of long-span bridge traffic loading explained in this chapter will inform the next generation of design and assessment practices, leading to continued safe service of existing bridges, and even better new designs.

Chapter 6

Factors affecting the accuracy of characteristic maximum load effects

*Eugene OBrien, Donya Hajializadeh,
Bernard Enright, and Cathal Leahy*

6.1 INTRODUCTION

There are many factors that affect the accuracy of characteristic maximum load effects (LEs). The process of fitting a distribution to the data using statistical Extreme Value theory was described in Chapter 3. The choice of methods, such as Block Maximum or Peaks Over Threshold, can significantly affect the result if the underlying assumptions of the theory are not respected. Tail fitting is often used, and the results from this fitting are clearly influenced by the point at which the tail is deemed to start. In some instances, the choice of Extreme Value distribution is also made a priori, which can further influence results. These statistical modelling aspects are reviewed in this chapter.

There are other challenges in highway bridge traffic loading that have received little attention. For example, the modelling of same-direction multi-lane traffic is a major challenge: it is inaccurate to combine statistical predictions from each lane because the distributions are not independent – a heavy vehicle in an overtaking lane is usually associated with a slow-moving vehicle in the adjacent slow lane that is probably also heavy. As seen in Chapter 3, a method known as Scenario Modelling has been developed to address this issue that is relatively easy to implement. In this chapter, a study on same-direction multi-lane traffic is reported. It examines the issues of correlations of vehicle weights between lanes and within lanes and how it affects characteristic maximum LEs.

Notional load models such as the AASHTO HL-93 and the Eurocode LM1, are normally used in the design of new bridges. Some bridges in a network will inevitably be subject to heavier vehicles than others, be constructed using different structural forms, materials, and to different standards, and be exposed to different environmental conditions. Therefore, bridges across a highway network can be expected to have different levels of safety. Clearly the notional load model must ensure an appropriate level of safety for the most heavily-loaded bridges, and so, as indicated in Chapter 3, notional load models are designed to envelope the predicted characteristic maximum LE values derived from the modelled traffic streams. However, a more difficult challenge is to develop a notional load model that provides

DOI: 10.1201/9780429318849-6

consistency in the level of safety for all LEs, in bridges of various structural forms, and for the full range of spans. An approach to achieve this goal of uniform levels of safety is addressed in Section 6.3.

Permit vehicles pose a significant challenge for bridge owners. The notional load model for bridge design is generally intended to represent extreme cases of standard, non-permit vehicles, and there are load models for extreme permit loading. Unfortunately, WIM systems do not generally have a mechanism for identifying which of the recorded vehicles have permits. Ideally, WIM stations would be equipped with a license plate recognition system that is linked to the permit database, but this is rare. As a result, notional load models for regular traffic can be biased by permit vehicles, and so it is desirable to filter them from the measurements. The imposition of an upper threshold based on weight does not work well as it creates a cluster of LEs close to the same value and distorts the statistical distribution of LE. An approach is described here that filters data based on the axle configuration. This is clearly an approximation but, in statistical terms, the process works quite well.

Bridge design and assessment loadings are intended to be valid for the bridge's design life. During that time, traffic growth can be reasonably expected to occur, and this is examined in Section 6.5. Freight traffic, measured in tonne-km, tends to grow with the value of the economy and is strongly correlated with Gross Domestic Product. Uncontrolled growth occurs in the form of increasing numbers of standard heavy trucks. Controlled growth also occurs, in the form of regulatory changes that allow heavier trucks, and special access networks.

WIM data is becoming ever more widely available and databases now exist with millions of vehicle records. However, for lightly-trafficked roads, such as secondary roads where traffic volumes are low, even when years of data are available, there may be very few examples of extreme vehicles or loading events. Section 6.6 presents a study where a large database, covering a number of sites, is used to enhance the data from a single lightly-trafficked site. This means that a more robust estimate of the characteristic maximum LE for a lightly-trafficked site can be determined.

6.2 CHOICE OF EXTRAPOLATION METHOD

6.2.1 The data

Traffic loading on bridges varies through time with many periods of zero LE when there is no traffic on the bridge. At other times, there are peaks of LE corresponding to heavy vehicle crossings or more complex vehicle meeting or overtaking events. Apart from fatigue issues, bridge owners/managers are generally interested in extreme values of traffic load effects as these have the greatest influence on an assessment of safety or reliability.

Extreme loading is usually determined with reference to a return period. For example, the AASHTO standard describes the characteristic maximum loading as that level which is only exceeded, on average, every 75 years. It is important to note that a return period is related to the level of safety and is independent of the bridge design life. Many statistical methods have been developed to determine characteristic maximum LEs. This section reviews these methods and discusses how the identified characteristic maximum can be influenced by the choice of method.

Finding characteristic maximum LEs is a process of statistical extrapolation or interpolation. As discussed in previous chapters, the data can be:

- Directly measured on a bridge of interest, for situations where there is a concern about a particular LE on that bridge;
- Calculated using measured vehicle weights and influence lines for the LEs of interest, often for a range of spans;
- Simulated using statistical properties found from measured vehicle weights. This can include more sophisticated loading scenarios than single crossings, such as meeting and overtaking events.

In the past, the data available did not usually extend to the return period of interest, and so a statistical extrapolation was done. With the computing power now available, long-run simulations can be carried out covering time periods up to and exceeding the return period. In these cases, fitting statistical distributions or models is a form of interpolation. While long-run simulation is only as good as the data and the assumptions inherent in that simulation, it removes the uncertainty around the curve fitting process. It is therefore a considerable improvement on the other approaches.

In the context of traffic loading, statistical modeling is a means of fitting distributions of maxima – for example, fitting the distribution of annual maximum bending moments. The most basic models rely on the 'iid' assumption, that is, that events are independent and identically distributed. Events are independent if one value is unrelated to and not caused by another. They are identically distributed when there are no changes to the distribution between events, due to traffic growth, for example. A complicating factor is that bridge traffic LEs are often made up from a mixture of iid distributions. As explained in Chapter 3, a bending moment for example, may be due to the crossing of a 5-axle truck, or a low loader with eight or more axles. The former is very common on most highways but its weight tends to be restricted by the legal limit. A low loader, on the other hand, may be much rarer, and be considerably heavier. If the bending moments due to both vehicle types are considered together, the statistical data will be a 'mixture distribution' and there can be a change in trend at some point. As ever more rare events are considered, the low loader LEs, and their statistical properties, will tend to become more prominent. It is important for accuracy that the fit to the tail does not incorporate data from more than one distribution.

6.2.2 Extreme Value theory methods

There are two main approaches to Extreme Value theory: Peaks-Over-Threshold (POT) and Block Maximum. Classical Extreme Value theory is based around the initial work by Fisher & Tippett (1928) and Gumbel (1935) and is the foundation for the Block Maximum approach. On the other hand, the POT approach was developed mainly by Pickands (1975).

POT considers data which exceed a specified threshold while the Block Maximum approach considers data that is the maximum in a specified block of time. An advantage of POT is that no extreme data is lost: if multiple extremes occur in the same block of time, POT considers all of them. In contrast the Block Maximum approach considers only the single maximum that occurs in that time frame. However, the choice of threshold in POT is subjective and more difficult to set than the choice of time block, which is simply governed by the need for stationarity, i.e. that there are no underlying cycles in the traffic stream with a period longer than the time block. So, because hourly flows clearly vary a lot through a day, the use of a block size of one hour is not recommended, while the use of a working day or a week (if variations within the week are to be considered) are valid block sizes. Clearly the choice of threshold or the choice of block size are critical to obtaining meaningful results.

6.2.2.1 Peaks-Over-Threshold

For the POT approach, LE values above the threshold are fitted to the Generalized Pareto distribution (GPD). Coles (2001) provides an outline proof that POT data converges to the GPD. A threshold as low as possible should be chosen, where the mean residual life plot becomes linear – see Coles (2001) for details. An early example of this approach in bridge traffic loading is the work by Crespo-Minguillón & Casas (1997) who select the optimal threshold based on a weighted least squares fit.

Having selected the POT threshold, the parameters of the distribution are estimated. There are several approaches to the estimation of these parameters. Bermudez & Kotz (2010) suggest the method of moments, the probability weighted method, the maximum likelihood method, and Bayesian updating. Maximum likelihood is a minimum variance estimator and can be recommended. However, the choice of estimator usually has less influence than the choice of threshold.

Estimation of the return level using POT must be done with reference to the data sampling interval. A return level is that level which is exceeded once in every, say, m observations, with each observation being recorded every

T minutes. For bridge traffic loading estimation, the calculated load effects are determined every timestep (e.g. 0.1 sec; see Chapter 3). Consequently, this extrapolation can become problematic since just one value of load effect per loading event is utilized, and so the average duration of a loading event is used as the time reference, *T*, which is uncertain.

6.2.2.2 Block Maximum

In the Block Maximum approach, the maximum LE in each block of time (day, year, etc.), is considered. As noted previously, any strong patterns or cycles should take place within the block duration. Typically, this means that hourly-maximum data is inappropriate due to hourly flow variations, but for some sites there may be strong weekly or even seasonal variations, requiring larger block sizes. Using time-based blocks has the advantage of time referencing the data which is convenient when calculating the characteristic LE for a specified return period. As explained in Coles (2001), block maximum data statistically converges to the Generalized Extreme Value (GEV) family of distributions which incorporates the classical Gumbel, Weibull, and Fréchet distributions (also known as Types I, II, and III Extreme Value distributions).

In this approach, in contrast with POT, there is a risk that some important data is discarded: if two unrelated extreme loading events occur in the same block of time, only one of the resulting LEs is retained. However, it is possible to expand consideration of the block maximum method to take account of more than one extreme values within a block (Coles, 2001).

In bridge traffic loading, most researchers explicitly fit the block maximum LE data to one of the Extreme Value distributions described by the GEV equation: Gumbel, Fréchet, or Weibull. The three types of distribution have distinct forms of behavior in the tail (Coles 2001). Many studies (Caprani & OBrien 2006, Caprani et al. 2008, Kanda & Ellingwood 1991, O'Connor & OBrien 2005) suggest that LE data is 'bounded,' i.e. it curves upwards towards an asymptote when plotted on Gumbel probability paper. The argument is that there is a physical upper limit to how much load can fit on a bridge, stemming from the capacities of the axle suspension systems. As a result of this physical limit, the Weibull distribution is most appropriate, for which the GEV shape parameter, $\xi < 0$. As Gumbel is a special case of Weibull, with parameter, $\xi = 0$, an assumption that LE is always in the form of the GEV equation, with $\xi \leq 0$, is recommended.

As for POT, different techniques have been used to fit distributions to block-maximum data. Grave (2001) uses a weighted least squares approach to fit Weibull distributions. Bailey (1996) describes the use of plots of the mean and standard deviation of load effects, to estimate the appropriate Extreme Value distribution. Moyo et al. (2003) plot daily maximum strain values on Gumbel probability paper and use a least-squares fit to determine the parameters of the distribution. However, as explained in Caprani (2005), the GEV is best fit using maximum likelihood as a minimum variance unbiased estimator.

6.2.2.3 Box-Cox-GEV distribution

The Box-Cox-GEV distribution is a more general Extreme Value approach and aims to address the disadvantages of both POT and GEV. Bali (2003) proposes the Box-Cox transform (Box & Cox 1964) to encompass the Generalized Pareto and GEV distributions (Caprani & OBrien 2009, Rocco 2010). This transformation may improve the rate of convergence to the limiting Extreme Value form since different distributions converge at different rates.

The parameters of the Box-Cox distribution are those of the GEV (μ, σ, ξ) plus a 'model parameter,' λ. As λ approaches unity, the distribution converges to the GEV and, as λ approaches zero, it converges to the GPD. To apply this model, a high threshold is set on the parent distribution (Caprani & OBrien 2009, Rocco 2010). Bali (2003) uses a threshold of two standard deviations about the sample mean. Caprani & OBrien (2009) consider thresholds in steps of 0.5 standard deviations in the range from -2.5 to +2.5 standard deviations about the sample mean. For the data examined, Caprani & OBrien (2009) find that highway bridge traffic loading is strongly in the GEV domain of attraction ($\lambda = 0.9$) but with statistical significance is neither GEV nor GPD solely. A drawback of the Box-Cox-GEV distribution is that the fitting is not robust due to the numerical complexity of the optimization function.

6.2.3 Tail fitting

6.2.3.1 Castillo's approach

Many sources of engineering data are from complex processes than cannot be confirmed to be independent and identically distributed, as required for the theoretical basis of Extreme Value theory to strictly apply. To address this, Castillo (1988) suggests that the top $2\sqrt{n}$ of a distribution of n data points should be used to fit the distribution – essentially a censored fit to the tail of interest. Enright (2010) suggests fitting to the top 30% of data, based on a sensitivity analysis. The assumption with this approach is that the fitted tail itself is composed of independent and identically distributed data, which is reasonable when the data is 'sufficiently extreme.' In bridge traffic loading this can be verified by confirming that the data is generated from loading events comprised of similar types of vehicle (e.g. low-loaders) and/ or number of vehicles (e.g. 3- or 4-vehicle multiple presences). While this approach is subjective and requires engineering judgment from an experienced practitioner, it can lead to simpler analysis that is meaningful.

6.2.3.2 Normal distribution

In the background studies for Eurocode 1, Flint & Jacob (1996) fit half-Normal curves to the upper tails of the histograms of LE. They adopt a least-squares best fit method to estimate the distribution parameters. Multimodal (bimodal or tri-modal) Gumbel and Normal distributions were also used.

For the calibration of the AASHTO load model, Nowak (1993) fits the load effects to a Normal distribution. The upper tails, corresponding to between three and four standard deviations, are used to project the results to about 5.3 standard deviations above the mean. These fits are then raised to an appropriate power to obtain the 75-year maximum LE distribution from which the bias factor and coefficients of variation – used in the reliability calibration – are determined. Like the first work for the Eurocode, this well-known early study was based on a very limited database. Just 9250 heavy vehicles were weighed, representing about two weeks of heavy traffic, on a highway in Ontario (Kulicki et al. 2007, Moses 2001, Nowak 1994, 1995, 1999, Nowak & Hong 1991, Nowak et al. 1993, Sivakumar et al. 2011). See Chapter 3 for more on this study.

6.2.3.3 Rice's formula

Rice's formula, introduced by Rice (1945) and described more recently by Leadbetter et al. (1983), is an approach for fitting the upper (or lower) tail of a Gaussian stochastic process that exceeds a certain threshold; i.e. it fits a parametric distribution to the histogram of 'upcrossings.' A key parameter in this approach is the number of bins in the upcrossing histogram. Cremona (2001) suggests the Kolmogorov test to select the optimal number of bins in the upcrossing rate histogram and the threshold. Cremona's approach is adopted by Getachew (2003) for the analysis of traffic LEs on bridges. O'Connor & OBrien (2005) compare the extremes predicted by fits to Rice's formula, Gumbel, and Weibull Extreme Value distributions. For bending moment in simply supported bridges for a range of span lengths, they find about 10% difference between the Rice formula and the other fits. Jacob (1991) uses Rice's formula to predict characteristic LEs for the cases of free and congested traffic in background studies for the development of the Eurocode.

6.2.4 Distribution of the prediction

In the methods described, there remains the compounding problem of parameter variability. That is, repeats of the same experiments will result in different parameter sets, even though the underlying random process may be the same. Some authors have proposed methods of capturing this influence of parameter variability on the predicted extreme.

6.2.4.1 Bayesian Updating

In Bayesian Updating, a distribution known as the prior, is first assumed. Then this prior distribution is updated as LE data becomes available (OBrien et al. 2015). The resulting posterior distribution reflects the evidence provided by the measurements. In this approach the predictive distribution takes account of the uncertainty in the parameter values, as measured by the posterior distribution. Lingappaiah (1984) develops bounds for the predictive probabilities of extreme order statistics under a sequential sampling scheme, when the population is either Exponential or Pareto. From a practical viewpoint, the most important issues arising from the Bayesian literature are the elicitation and formulation of genuine prior information in Extreme Value problems, and the consequent impact such a specification has on inferences. Coles & Tawn (1996) consider a case study in which expert knowledge is sought and formulated into prior information as the basis for Bayesian analysis of extreme rainfall.

6.2.4.2 Predictive Likelihood

Predictive Likelihood is a frequentist (i.e. non-Bayesian) approach to allowing for parameter uncertainty in extrapolations. It is proposed for bridge traffic loading by Caprani & OBrien (2010). Referring to the predicted value as the 'predictand,' the joint likelihood is calculated of observing both the actual data and a postulated predictand. This is repeated for a range of possible predictands, and the resulting distribution of predictive likelihoods is normalized to find a distribution of predictands. Predictive Likelihood can therefore be viewed as a method which 'ranks' possible realizations of the predictand according to how likely they are to occur, given the data. This 'ranking' results in the predictive distribution. To include the parameter variability within each fitting, a Modified Predictive Likelihood approach can be used (Barndorff-Neilsen 1983, Butler 1986). However, Caprani (2005) finds that the allowance for parameter variability only slightly affects the final predictive distribution for the bridge traffic loading studied.

6.2.5 Comparative study of extrapolation methods

6.2.5.1 Simple Extreme Value problem

OBrien et al. (2015) explore the nature of the extrapolation problem with a simple numerical example, with a known theoretical answer, as a benchmark case study to compare the methods of extrapolation described above. In the next section, this is extended to more realistic problems, where long-run simulations are used to provide the benchmark results, against which each method is compared.

The probability of failure is the most definitive measure of bridge safety. However, it is strongly influenced by the resistance model which varies greatly from one example to the next. Therefore, to retain the focus on load effect, the comparisons made here between the approaches are conducted using characteristic maximum values.

For the theoretical example, a Normally distributed random variable (such as vehicle weight in tonnes) is considered. The random variable, W, is Normally distributed with mean 40 and standard deviation of 5, i.e. $W \sim N(40, 5)$. As an example, $n = 3000$ values of W in a given block, considered to be one day, with the maximum-per-day random variable, $X = \max (W_i)$ where $i = 1,2,...,n$.

In practice, a finite number of days of data is available and Extreme Value distributions are inferred from a dataset of, for example, daily maximum values. These daily maxima (X values) may be used to infer the distribution of annual maximum load effect, say Y. The days are generally considered to be working days and a year is usually taken to consist of 250 such days (allowing for weekends and about 11 public holidays, when economic activity is less). The exact theoretical solution to this problem can be calculated from its cumulative distribution function, as: $F_Y(w) = \left[F_X(w) \right]^{250} = \left[F_W(w) \right]^N$, where N is the number of values in a year, equal to $(3000 \times 250 =)$ 750,000. It is worth noting here that the distribution of maxima from the Normal distribution is slow to converge to the Gumbel distribution (which is its theoretical limiting distribution), and so the exact distribution is used instead.

The extrapolation approaches are tested to estimate the characteristic maximum value for a 75-year return period. Four methods are tested first: POT, GEV, Box-Cox-GEV, and Normal. For the Extreme Value and Normal distributions, a least squares fit is found for the top 30% of values from 1000 daily maximum LEs. Twenty samples are generated using the same parameters to investigate the variability associated with the results (i.e. each time, a new sample of data is generated and the same calculation is performed to find the characteristic value). The results for the four methods are illustrated in Figure 6.1.

The 20 randomly generated samples exhibit different types of tail behavior – Gumbel, Fréchet, and Weibull – and vary around the exact distribution. POT, GEV, and Box-Cox all give similar results, with characteristic values varying within the range -5% to +10%. The Normal distribution fits vary less than the others but there is a small non-conservative bias in the calculated 75-year values.

The other three methods of extrapolation are illustrated in Figure 6.2. While Cremona (2001) considered a variable quantity of data for the Rice Formula, the top 30% is used here to provide a direct comparison with the other tail fitting methods. Hence, the optimized parameters are found using a best fit to the normalized upcrossing histogram for the top 30% of daily maximum data.

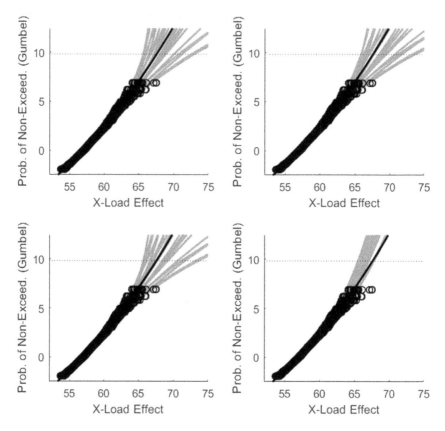

Figure 6.1 Twenty probability paper plots, each for 1000 maximum-per-day load effects, with fits in gray and exact plot in black; (a) Peaks-Over-Threshold; (b) Generalized Extreme Value; (c) Box-Cox-GEV; (d) Normal. Note: a 75-year return period corresponds to a value of 9.8 on the vertical axis.

Unlike the tail fitting methods, all daily maximum LEs are used for Bayesian Updating. The data is assumed to be GEV with uncertain parameter values. In effect, a family of GEV distributions is considered. These parameter values are initially assumed to be equally probable within specified ranges (uniform prior distributions). The daily maximum data is then used to update their probabilities.

Predictive Likelihood is also based on the entire dataset of block maximum values fitted to a GEV distribution. The joint likelihood is calculated for a range of possible values at a given level of probability (predictands), given the value of that predictand and the available daily maxima.

Figure 6.3 provides a summary of the results in the form of mean values for the 20 runs in each case, ± one standard deviation. In addition to the 1000 day samples, corresponding results are presented for smaller sample sizes – 200 and 500 days – to show how it affects the variability of results.

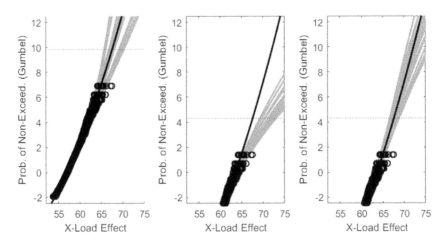

Figure 6.2 Probability paper plots for whole distribution fitting methods (in gray), with exact plot in black: (a) Rice formula; (b) Bayesian Updating; (c) Predictive Likelihood.

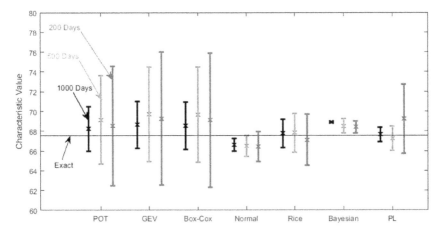

Figure 6.3 Inferred 75-year characteristic values and error bars (mean ± one standard deviation) (PL - Predictive Likelihood).

In this figure, the horizontal line corresponds to the exact characteristic maximum value. Interestingly, there is not a strong trend in the influence of the quantity of data on the mean result: going from 200 to 500 days of data sometimes results in a fall in accuracy though the 1000 day results are generally more accurate. The variability does reduce, as would be expected, when more data points are available.

The exercise shows that most methods are somewhat biased conservatively (i.e. tend to be greater than, or similar to, the exact solution). The extreme-value approaches exhibit considerable variation in predictions, especially for the smallest data sample when they are about ±10%. This is likely because of the fitting to a Fréchet-type of distribution and could be considerably improved if this behavior were blocked. Fitting to a Normal distribution performs well, likely for the same reason. Rice's formula, Bayesian Updating, and Predictive Likelihood are generally quite good, though it should be noted that the low variability in repeat runs is clearly not an indication of accuracy. Bayesian Updating, in particular, gives a very narrow range of results though the mean ± one standard deviation does not enclose the exact value. However, it should be noted that this example was based on a single iid distribution. In a more realistic mixture distribution, with LEs coming from different types of vehicle for example, methods that use all the data will be less effective than those based on a tail fitting approach.

These results can be compared in different ways. For example, Predictive Likelihood and Rice's formula for 1000 days of data are the best if the assessment criterion is the deviation of the mean characteristic value from the exact result. The other methods are at similar, lesser levels of accuracy. For small quantities of data, no single method stands out as the best. It is of interest that Predictive Likelihood is now less accurate than several of the other methods.

6.2.5.2 Traffic load effect problem

The second set of examples used by OBrien et al. (2015) is based on a carefully calibrated traffic load simulation model. This study addresses the issue of whether or not the data should be constrained from Fréchet-type behavior. A series of simulations is used, along with a long-run benchmark simulation. The load effects considered for this exercise are summarized in Table 6.1.

As part of the European 7th Framework *ARCHES* project [1], extensive WIM measurements were collected at five European sites: in the Netherlands, Slovakia, the Czech Republic, Slovenia, and Poland. The *ARCHES* site in Slovakia is used as the basis for the simulation model presented by OBrien et al. (2015). Measurements were collected at this site for 750,000 trucks over 19 months in 2005 and 2006. The traffic is bidirectional, with average daily truck traffic (ADTT) of 1100 in each direction. A detailed description of the methodology adopted is given by Enright & OBrien (2013).

Table 6.1 Load Effects and Bridge Lengths

	Load Effect	Bridge Lengths (m)
LE1	Mid-span bending moment, simply supported bridge	15, 35
LE2	Shear force at start/end of a simply supported bridge	15, 35
LE3	Central support hogging moment, 2-span continuous bridge	35

For the benchmark 5000-year simulation, the outputs consist of annual maximum LEs. The 5000 annual maximum data points are then used to interpolate the 75-year maximum mean value to a high degree of accuracy. This long-run simulation process is considered to be highly accurate, subject to the assumptions inherent in the model, and is used as the benchmark against which the accuracy of all other methods is measured. It should be noted that the results for long-run simulation are strongly site-specific. The load effects considered for this exercise are summarized in Table 6.1.

In this exercise, simulated measurements representing 1000 daily maxima, are used as the basis for extrapolation to estimate the 75-year maximum, and so are compared with the benchmark results. Twenty repetitions are conducted. The probability paper plots are shown in Figure 6.4. Some scatter in the results is evident in the 20 repeat runs, as can be expected. For some load effects, particularly support shear, the distribution of the data is multi-modal, i.e. there is a change in slope in the probability paper plot (Figure 6.4(b)). This implies data from a mixture of different parent distributions. This could be due to, for example, the change from (i) daily maxima arising from heavily loaded regular trucks to (ii) maxima arising from extremely heavy (and rarer) low loader vehicles and cranes. As a result of this inhomogeneity, as discussed previously, for the Extreme Value and tail fitting methods, the distributions are fitted to the top 30% of data.

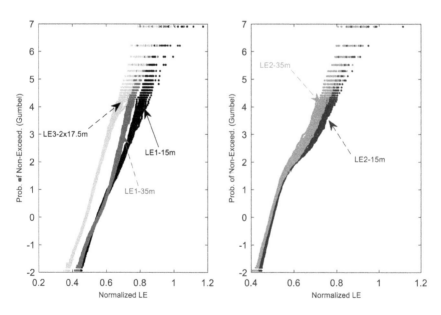

Figure 6.4 Probability paper plots, normalized with respect to the benchmark 75-year characteristic maximum value: (a) mid-span moment in simply supported, LE1 and central support moment in 2-span, LE3; (b) support shear in simply supported, LE2.

The results are presented in Figure 6.5. As for the simple example, it can be seen that the Extreme Value methods are capable of covering the benchmark characteristic value within one standard deviation in most cases, but that range is considerable: from about 90% to 120% of the benchmark. The exception to this is LE3, the central support hogging moment in a 2-span bridge. The influence line for this load effect has two peaks and is quite sensitive to the axle configuration of the vehicle. It is probably for this reason that the samples of 1000 daily maxima are highly variable, with different samples giving quite different results for the first three tail fitting

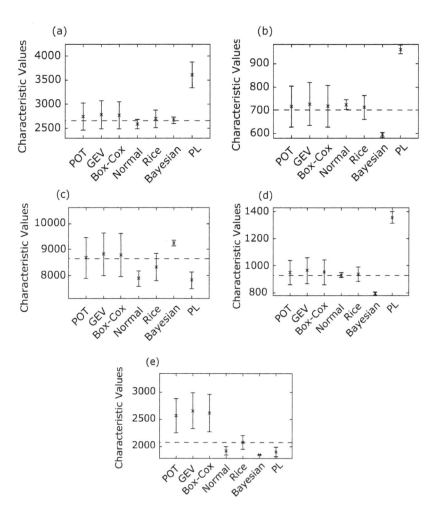

Figure 6.5 Inferred 75-year characteristic values from 1000 days of data (mean ± one standard deviation): (a) LE1, mid-span moment, 15 m; (b) LE2, support shear, 15 m; (c) LE1, mid-span moment, 35 m; (d) LE2, support shear, 35 m; (e) LE3, central support moment, 2x17.5 m spans.

methods. The results are consistently conservative, but the accuracy is poor, with the range (mean ± one standard deviation) covering about 40% of the benchmark value for this LE.

The fit to a Normal distribution performs better than the Extreme Value methods. This is probably due to Fréchet-type behavior in the GEV fits which is inconsistent with the nature of the data. By not allowing Fréchet-type behavior, the Normal fits are generally much better, and consistently so for the 20 repeat runs.

The Bayesian approach gives different ranges of accuracy for different load effects. Several variations were tested in attempts to find a Bayesian approach that is consistently good. The results shown are based on the use of all the data; updating just two parameters (σ and ξ), with a uniform prior distribution and a restriction on the range of ξ to be non-positive, i.e. to prevent Fréchet-type behavior. As a result, it is a somewhat unfair comparison with the other approaches.

Predictive Likelihood performs much less well than in the simpler example. This may be because the data resulting from long-run simulations comes from a mixture distribution rather than a single distribution. The tail region in the data fits well to a single distribution but not to the entire data set. As a result, any method that takes account of all data and tends to fit a single distribution to it, tends not to provide good results relative to the benchmark. It should be noted that this performance is quite different to that found by Caprani & OBrien (2010) for mixed traffic.

6.2.6 Extrapolation recommendations

In summary the prediction of characteristic value is more influenced by the database being used than the method adopted. Indeed, several studies have shown that truck traffic is highly dependent on the site characteristics (Nowak & Rakoczy 2012, Zhao & Tabatabai 2012, Sivakumar et al. 2007, Pelphrey et al. 2008). Kim and Nowak (1997) also highlight the site dependence of characteristic traffic load effects, even within a geographical region. Further, increasing the number of data in the sample tends to result in higher accuracy of approximation but it does not eliminate the uncertainty associated with the extrapolation process. Some of the main points that can be taken from this study are:

- In general, POT, GEV and Box-Cox-GEV give similar, reasonably good, results for most LEs;
- While it was not explicitly tested in this study, it appears that a significant improvement in accuracy can be achieved by enforcing non Fréchet-type behavior, particularly for some LEs and spans. This can be achieved by fitting to a Normal distribution or by fitting to GEV with a requirement that the shape parameter, ξ, be non-positive;

- Rice's formula performs very well, generally better than other methods;
- Bayesian Updating can also give good results, but the accuracy is inconsistent, unpredictable, and requires subjective tuning;
- The variability in results when the method is repeated multiple times is not a reflection of the accuracy, i.e. the results can be biased;
- Predictive Likelihood does not perform well when applied as a single distribution to a mixture population;
- Central support moment in the 35 m long 2-span bridge has been found to be a particularly challenging LE.

Since traffic load effects result from several statistical distributions, there is no one approach that works well in all situations. Indeed, it would seem that the choice of method is unimportant compared to some of the other choices being made in the process. Most of the methods give results within about ±10%, most of the time. If more accurate results are needed, long-run simulations are an excellent means of avoiding the uncertainties around these extrapolation processes. For situations where the distribution must be known precisely, it is recommended to conduct bespoke studies for the site and traffic under consideration.

6.3 THE NATURE OF EXTREME TRAFFIC LOADING EVENTS

6.3.1 Correlations in same-direction multi-lane traffic

In bridges with two opposing-direction traffic lanes, the weights of vehicles in different lanes can be reasonably considered to be statistically independent. For adjacent same-direction lanes, however, the weights have been found to be weakly correlated. Even within a lane, there is a small correlation in vehicle weights due to the daily traffic patterns. For example, if a heavy truck occurs, then some studies have found a higher probability that the next truck will also be heavy. This is partly because there is an increased probability that it is the time of day when heavy trucks tend to travel.

In early work by Nowak (1993), a number of simplifying assumptions are made on truck weight correlations – for example that one in 15 heavy trucks has another truck side-by-side, and that for one in 30 of these multiple truck events, the two trucks have perfectly correlated weights. Kulicki et al. (2007) note that these assumptions were based on limited observations, and the assumptions on weight correlation were entirely based on judgment, as almost no data were available. Moses (2001) estimates multiple presence probabilities as a function of average daily truck traffic (ADTT), and selects conservative values, some based on subjective field observations. Sivakumar et al. (2007) define side-by-side events as two trucks with a headway separation of ± 18.3 m (60 ft) and consider the influence of the bridge length. Sivakumar et al. (2011), citing Gindy and Nassif (2006), refine this model by classifying multiple-presence events as side-by-side, staggered, following or multiple. They extract statistics from WIM data, on the frequency of occurrence of these events for different truck traffic volumes and bridge spans. However, there is no allowance for correlation between the weights of trucks in adjacent lanes.

Enright uses WIM data from sites in the Netherlands and the Czech Republic (OBrien & Enright 2011, Enright 2010) to investigate the significance of correlations in truck weights and suggests that it may account for an increase in characteristic maximum load effect of up to 8%. He reveals that, for groups of adjacent vehicles in both lanes, there are patterns of correlation and interdependence between vehicle weights, speeds, and inter-vehicle gaps. For both sites, there are significant differences between the two lanes, with a much higher proportion of light vehicles in the fast lane. Here it is worth noting that European countries typically have asymmetric passing rules, i.e. it is only allowed to overtake on one side. Hence, this finding may not apply readily to countries with symmetric passing rules (e.g. the United States).

Within lanes, there are correlations between a leading truck and the truck immediately following it, the second truck behind it and so on. As shown in Figure 6.6, the correlation with the first following truck is almost 5%, with the second following truck is 2.5 to 3%, and there is an underlying correlation with all further following trucks of about 2%. The greater correlation with nearby trucks may be due to driver behavior where groups of associated vehicles travel together. Similar patterns are evident in the fast (overtaking) lane, with an underlying level of correlation of 7.4% and a pairwise correlation of 9.4% in the Netherlands. The corresponding figures

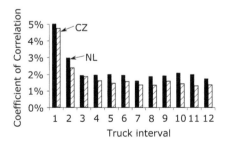

Figure 6.6 Autocorrelation of gross vehicle weight, slow lane (after Enright 2010) for two European sites (CZ – Czech Republic; NL – Netherlands).

for the Czech Republic are lower, at 1.4% and 2.2% respectively. The cause of this difference is not readily apparent.

For short- to medium-span bridges, loading events featuring one truck in each same-direction lane, either side-by-side or staggered, are important. To assess if there is any dependence between the weights of these vehicles, Enright (2010) paired each fast-lane truck in the measured data with the nearest truck in the slow lane, and recorded the inter-vehicle gap between the two vehicles. At both sites, most fast-lane trucks are within two seconds of a slow-lane truck – 75% in the Netherlands (see Figure 6.7(a)) and 72% in the Czech Republic.

The average GVW of the truck in the fast lane and of the nearest truck in the slow lane are plotted against the inter-lane gap in Figure 6.7(b). There is a significant peak in average fast lane GVW when gaps are around zero, i.e. when the trucks are overtaking, and a similar pattern is evident in the Czech Republic. Thus, a heavy truck in the fast lane seems likely to be associated with a nearby truck in the slow lane, i.e. it is overtaking a slow-lane truck. This is highly significant as it means that in two-lane loading events, the fast-lane truck, being correlated, should be heavier than the regular WIM data suggests.

6.3.1.1 Simulating trains of truck traffic

Measured WIM data rarely provides enough LEs for direct fitting of a statistical distribution and extrapolation to find the characteristic maximum. It is therefore common practice to extend the population of LEs by simulating many more vehicle crossing and overtaking events than are present in the WIM database. Clearly, for most accuracy in the simulation of trains of traffic, the observed vehicle weight correlations should be included. In multi-lane traffic, this becomes challenging. For two lanes of traffic, three gap distributions exist: the in-lane gaps for each of the two lanes and the inter-lane gaps. Monte Carlo simulation can generate typical weights for each lane of traffic, which allows for the in-lane correlations, but then the

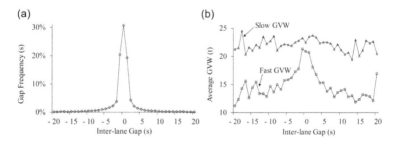

Figure 6.7 Relationships between vehicles in adjacent same-direction lanes, the Netherlands: (a) frequency of inter-lane gaps; (b) gross vehicle weights by inter-lane gap.

inter-lane correlation will be lost. Alternatively, the slow lane traffic, with its correlations, can be simulated and the inter-lane correlation, but then the fast lane correlations will be incorrect.

The Scenario Modelling approach developed by Enright & OBrien (2011) (see Chapter 3) allows for the implicit correlations of multiple parameters in multi-lane traffic. Enright (2010) uses this approach, with long-run simulation, to generate typical characteristic maximum loading scenarios for 2-lane traffic. He reviews the top 20 loading scenarios in 1000 years of simulation. The proportion that involves two trucks in these critical loading events depends on the lateral distribution (lane factor) of the bridge. Enright defines the lane factor as the contribution of the second lane traffic to the load effect, expressed as a percentage of the contribution of the first lane traffic. Consequently, this depends on the location of the girder of interest – a girder located where two lanes meet will have a high lane factor. It also depends on the transverse stiffness of the bridge. For example, a beam-and-slab bridge is transversely flexible and will tend to have a low lane factor. Where bridges have a high lane factor, there is clearly a greater likelihood that 2-truck loading events will govern. For bending moment in bridges with high lane factors from 15 to 45 m long, 60% to 100% of the critical loading events involve two trucks. Within these, the average contribution of the fast lane truck ranges from 16% to 43%. The figures are much less for shear force, with fast lane trucks typically governing in less than one third of cases and then contributing less than 10%. For bridges that are not stiff transversely, the number of critical 2-truck loading events is much less and, when they do occur, the contribution of the fast lane truck is generally less than 10%.

Some 1000-year characteristic maximum LEs from Enright's (2010) study are illustrated in Figure 6.8 and provide insights into what might govern in a 1000-year return period. For a bridge with a low lane factor, single truck loading events tend to govern but not always. Figure 6.8(a) shows two critical loading events for mid-span moment in a very short bridge subject to Dutch WIM data. A 191 tonne low loader on 14 axles governs in one simulation and a 175 tonne low loader being overtaken by a 50 tonne 6-axle truck in the other. Low loaders are less common for the Czech site and, for the same LE, the governing scenarios are a single 127 tonne truck and a 124 tonne truck being overtaken by a 29 tonne 5-axle truck.

On longer bridges, longer vehicles are likely to emerge as critical contributors to the extreme loading events. Single-vehicle loading events govern in all four cases shown of Figure 6.8(b) which is for support shear force in a 25 m simply supported bridge with a low lane factor. For a longer bridge with a high lane factor, more 2-vehicle loading events are likely to feature which is the case in Figures 6.8(c) and (d). It is of interest that, in the two-vehicle loading events, the fast lane vehicle tends not to be particularly heavy.

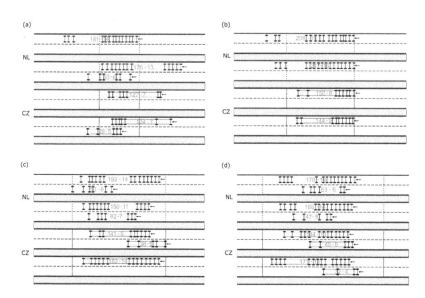

Figure 6.8 Typical 1000-year characteristic maximum loading events, two each using the Netherlands (NL) and Czech Republic (CZ) traffic (after Enright 2010). (The numbers, #–# indicate the vehicle weight in tonnes and the number of axles respectively.) (a) Mid-span moment, 15 m simply supported bridge with low lane factor; (b) Shear at support of 25 m simply supported bridge, low lane factor; (c) Hogging moment over central support of 35 m long 2-span bridge, high lane factor; (d) Mid-span moment in 45 m simply supported bridge, high lane factor.

6.3.2 Consistency of safety in notional load models

For the assessment of existing bridges, site specific characteristic maximum LEs can be calculated using the methods described here. However, this process is too complex for the design of new bridges and codes of practice specify notional load models. As explained in Chapter 3, these are generally intended to generate similar magnitudes of LE to the characteristic maximum values. The notional load model may include a representation of a real vehicle, as is the case for the AASHTO standard (Figure 1.3), or it may simply consist of point loads and uniform loading, as is the case for the Eurocode (Figure 1.5). As noted in Chapter 3, notional load models are, by their nature, compromises between a wide range of LEs, bridge types and spans. As a result, bridges designed using them can have considerably different levels of safety considering different traffic regimes and, even within one bridge, there are different levels of safety for different LEs.

In this section, the AASHTO (2012) HL-93 notional load model is examined against WIM measurements. The HL-93 notional load model is intended to generate 75-year return period characteristic load effects for normal vehicular use of the bridge (AASHTO 2012). NCHRP Report 683 (Sivakumar et al. 2011) interprets this as all legal trucks, illegal overloads, and un-analyzed permits (all routine permits) as they represent normal service traffic. This is referred to in the specifications as 'Strength 1' loading.

Leahy et al. (2014) investigate the differences in safety levels between bridge LEs using an American WIM database when compared with the safety implied by the HL 93 notional load model. The database contains 74 million trucks from 17 sites in 16 American states. It was collected in the period 2005–2011 under guidelines to ensure its 'research-quality' (Walker & Cebon 2012, Walker et al. 2012). Strength 1 vehicles are filtered using state-specific permit and weight regulations, as recommended by NCHRP Report 683 (Sivakumar et al. 2011).

6.3.2.1 HL-93 – Single-lane bridges

For a single-lane bridge, three load effects are examined on 20, 30, 40 and 50 m total bridge lengths: mid-span bending moment on a simply supported bridge (LE1), shear at the end support on a simply supported bridge (LE2) and hogging moment over the central support of a two-span continuous bridge (LE3). In each case, the 75-year characteristic maximum values are obtained by extrapolation from the observed load effect values. They are then normalized with respect to the corresponding HL-93 load effects and illustrated in Figure 6.9. The sites are ordered according to the 'severity' of loading, i.e. by how much, on average, the load effects exceed the corresponding HL-93 value.

The average normalized load effects across all sites is 1.062, i.e. the true load effects are greater than the corresponding nominal HL-93 values. It is important to note that the characteristic LE values are far below 1.75, corresponding to the factored HL-93 LE for Strength 1 loading. A significant point to note is the spread of the values at each site. For example, 50 m two-span continuous bridges are being over-designed for LE3 at most sites while most of the corresponding 30 m bridges are being under-designed. For the over-designed load effects, this results in excess capacity and a waste of material, whereas for the under-designed load effects, the bridges may be below the target level of safety.

6.3.2.2 HL-93 – Two-lane-bridges

Unfortunately, there were far fewer sites in the WIM database with quality data available for more than one lane, and so Leahy et al. (2014) present 2-lane loading results for just three sites, one of which includes data for both directions. In total the data comprises 8.4 million truck records and almost 15 site-years. Scenario Modeling is used to generate trains of 2-lane traffic and simulate 300 years of this traffic to determine the 75-year

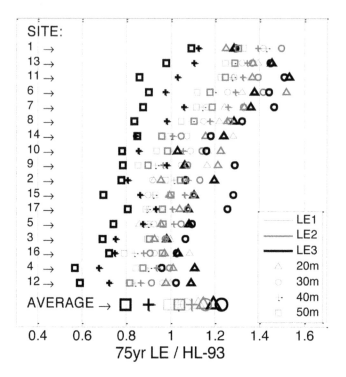

Figure 6.9 Single-lane characteristic load effects, normalized with respect to HL-93 (adapted from Leahy et al. 2014, with permission from ASCE).

characteristic maximum values. This process is repeated for four spans (10, 15, 20, and 30 m), two bridge cross-sections (solid slab and girder), the four sites/directions, and two transverse locations of interest (edge of slow lane and center between lanes). It can be seen in the results, Figure 6.10, that the HL-93 model has a higher level of conservatism for center load effects (in black, generally to the left) than for edge effects (in gray, to the right), with about 10% difference. This is unsurprising as the HL-93 model represents a loading event with the same load in both lanes, whereas the long run simulation results show that the critical side-by-side events typically involve a heavy slow lane truck being overtaken by a lighter fast lane truck. The different levels of conservatism for edge and center effects, and the simulation results, suggest that a load model should have a larger load in the slow lane than the fast lane. It is also seen that the model is less conservative for the shorter slab bridges. Of course, these findings are based on data from just four WIM sites/directions and cannot be applied to all sites. Nevertheless, it is clear that there can be quite different levels of safety implied by a single notional load model and this may be of significance in a special assessment or design situation.

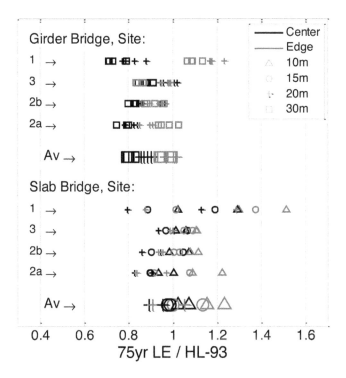

Figure 6.10 Two-lane same-direction characteristic load effects, normalized with respect to HL-93 (adapted from Leahy et al. 2014, with permission from ASCE).

6.4 SEPARATING STANDARD FROM NON-STANDARD TRUCKS

For the design of notional load models (see Chapter 3) for standard vehicular traffic – those not requiring a special permit – the WIM data forming the basis of the model should not contain special permit vehicles. Since WIM measurements include all vehicles, those special permit vehicles ought to be filtered, prior to using the data for assessment or notional load model development. Whether or not a vehicle has a permit is a complicated question: simply imposing the legal weight limits on the data removes some of the most significant data, namely the illegally overloaded standard vehicles, and in the process, distorts the statistical model.

As noted above, in the United States standard vehicles are interpreted as legal trucks, illegal overloads, and routine permits (Sivakumar et al. 2011) – see Figure 6.11. 'Grandfathered rights' trucks are those allowed to exceed the legal limit in some states because they were permitted to do so before the limit was introduced. Similar arrangements may exist in other countries. A routine permit is one that, for example, may be issued annually and

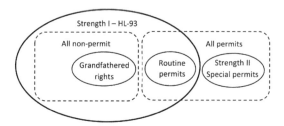

Figure 6.11 Vehicle categories in the United States (after Leahy et al. 2014).

that allows the vehicle to travel unescorted at full highway speed. Thus, regular legally loaded trucks, illegally overloaded trucks, trucks with grandfathered rights, and trucks with routine permits are all considered to be standard trucks and constitute 'normal' traffic for bridge loading purposes. Interestingly, the Eurocode defines 'abnormal load' as one which may not be carried on a route without permission from the relevant authority, suggesting that 'normal' load consists of non-permit vehicles. It does not mention routine permits but it seems reasonable to include these as part of normal loading. Thus, the terminology 'standard trucks' will be used here to refer to normal loading (including routine permit trucks) and 'non-standard' to all others. Many authors (Nowak 1993, Bailey & Bez 1999, Enright & OBrien 2013) make no distinction between standard and non-standard trucks and calculate characteristic maximum LEs using all vehicles in the WIM database. However, non-standard vehicles are controlled, at least to some extent (Luskin et al. 2000, Koniditsiotis et al. 2012), and bridge safety assessments can consider standard and non-standard trucks using separate statistical models, reflecting their differing variabilities.

6.4.1 Removing apparent non-standard vehicles from WIM data

Enright et al. (2016) propose a filter to separate the truck population into two types: 'apparent standard' and 'apparent non-standard.' Both truck types are identified based on their axle layout, rather than on their weights, as trucks of either type may be illegally overloaded. They acknowledge that the apparent standard group may contain trucks with the axle configuration of a standard truck but which have a permit to carry increased weight, and similarly, that the apparent non-standard group may contain trucks which do not have a permit and are travelling illegally. Nevertheless, this approach classifies the truck population into two statistically distinct subpopulations.

For the study, Enright et al. (2016) analyze 2.7 million truck records from five European countries and 81.6 million records from 17 American states (see Section 6.3). A series of rules, based on axle spacings only, are used to classify the vehicles. The rules vary between the US and Europe and some are subjective. They classify European non-standard trucks into three types: (i) low loaders, (ii) mobile cranes, and (iii) trucks carrying crane ballast – see Figure 6.12. 'Low loaders' consist of a tractor and trailer and have one large inter-axle spacing, usually in the range 8–13 m. 'Mobile cranes' have a rigid body and consist of closely spaced axles with relatively large axle loads. 'Crane ballast trucks' consist of tractor and trailer units. Their axles are generally closely spaced except for one slightly larger spacing between the tractor and the trailer.

American non-standard vehicles are a little different. Again, three types can be identified but this time they are: (i) low loaders, generally longer and with more axles than in Europe, (ii) mobile cranes, generally lighter and with fewer axles than in Europe, and (iii) mobile cranes with dollies, as illustrated in Figure 6.13. The trend in the US towards carrying large weights on longer vehicles may be because of the federal bridge formula (Jacob et al. 2010). Figure 6.14 shows typical low loaders from the two databases – the total weights are similar, but the US vehicle is longer. The mobile cranes in the US generally have fewer axles than Europe, probably because the heavier mobile cranes rest the boom on a trailing dolly to allow the weight to be spread over a greater length (Figure 6.13). The crane ballast trucks in the European WIM data were not found in the US data. It is assumed that ballast is carried in multiple smaller trucks. An increasingly common feature of European traffic is the European Modular System (EMS) truck which is allowed in a number of countries (Akerman & Jonsson 2007). These can have up to nine axles and can be up to 25.25 m long but are considered here to be standard.

Figure 6.12 Non-standard trucks from a WIM site in the Netherlands: (a) low loader; (b) mobile crane; (c) crane ballast truck (after Enright et al. 2016).

Figure 6.13 US non-standard vehicle type 3: mobile crane with dolly.

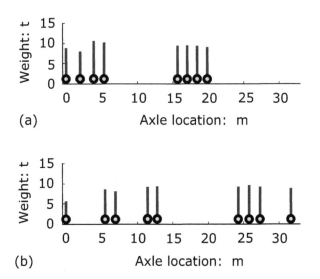

Figure 6.14 Axle configurations of low loaders: (a) European low loader with 75 t on eight axles; (b) longer American low loader with 79 t on nine axles (note that the heights of the vertical lines are the axle weights in t.) (after Enright et al. 2016).

Figure 6.15 shows a probability paper plot of maximum-per-day GVWs from an American and a European test site, filtered using these rules. The trend in the apparent non-standard vehicle data is smooth for the Netherlands data. This suggests that the maximum-per-day weights are identically distributed. Given that the data is from a mixture of vehicle types (low loaders and crane-type vehicles), this is somewhat surprising. Mobile cranes in Europe tend not to exceed about 110 t and the maximum-per-day weight on the majority of days is from a low loader. The distribution in the American non-standard vehicle weights is less smooth, suggesting that different vehicle types are governing on different days. This curve is also more bounded, i.e. it tends to curve upwards towards an asymptote, suggesting a physical upper limit. This may be the result of truck permitting regulations in Arizona. Unsurprisingly, the characteristic maximum GVWs for apparent standard vehicles are considerably less in both countries than for apparent non-standard.

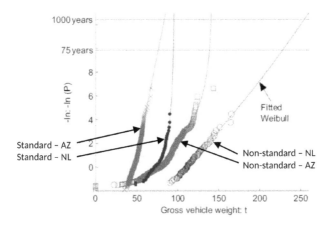

Figure 6.15 Probability paper plot (Gumbel scale) of the gross weights of apparent standard and apparent non-standard vehicles from Arizona (AZ), USA and the Netherlands (NL) (after Enright et al. 2016).

6.4.2 Modeling non-standard vehicles

It is generally considered that a truck would not be given a permit if it might subject the bridge to a dangerous overload. However, research in the US has found that bridge load effects caused by permit trucks regularly exceed the maximum effects expected by the permitting authority (Zhao & Tabatabai 2012). Moses (2001) gives the example of a 250 tonne truck that illegally travelled over 160 km in Ohio before being stopped by authorities. WIM evidence suggests a similar situation in Europe: non-standard vehicles sometimes breach the regulations for allowable weight and for the presence of an escort vehicle. This suggests that it is not safe to assume that bridge load effects caused by permit trucks will always be within the limits set by the permitting authority, and that a probabilistic analysis is necessary.

Enright et al. (2016) propose a method of modeling non-standard vehicles, addressing the challenge of there typically being very few records in most databases. They first separate the data into three types: low loaders, mobile cranes, and a third type that depends on the source of the data; crane ballast vehicles for Europe and mobile cranes with dollies for the US. They propose a simplified version of each vehicle type (Figure 6.16). For each vehicle type, GVW and the number of axles are first generated using a bivariate Normal distribution. Figure 6.17 shows that the most frequent records have seven or eight axles and weigh around 70 to 90 t. As the weights get higher, the number of axles also increases: it can be seen that 100 t vehicles are likely to have nine or more axles. The fit is quite approximate, but this is a product of the sparsity of the data. Axle spacings are modelled as shown in Figure 6.16, but for low loaders, the location and magnitude of the large gap is particularly important for the hogging

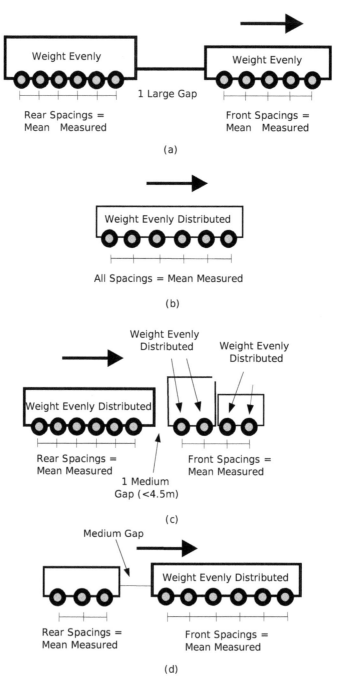

Figure 6.16 Simplified representations of the main non-standard vehicle types (after Enright et al. 2016): (a) low loader; (b) mobile crane; (c) crane ballast vehicle (Europe only); (d) mobile crane with dolly (US only).

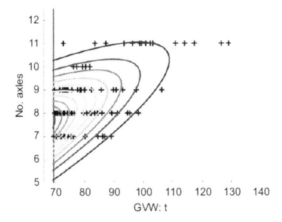

Figure 6.17 Bivariate Normal distribution fit to GVW and number of axles for low loaders recorded at the Czech Republic site.

moment at the central support of a 2-span continuous bridge. Normal distributions are fitted to (i) the magnitude of the gap, (ii) its location in the
vehicle, and (iii) the proportion of the GVW in the front versus the back
of the vehicle. The axle weights within each axle group are assumed to be
equal. For mobile cranes, the load is assumed to be equally distributed on
evenly-spaced axles. Crane ballast trucks are treated similarly to low loaders. However, there is a further complication as the two axles immediately
before the gap tend to be heavier, and a fourth parameter is used to generate
the relative magnitude of these. Finally, cranes with dollies are simulated in
a similar way to low loaders.

On the basis of these assumptions, a vehicle stream can be generated
and the resulting LEs calculated. A final complication is the tendency for
non-standard vehicles to travel in convoys. For example, at the Netherlands
site, 6.1% of non-standard vehicles are followed by another one. This may
be important for longer bridges. The results of these simulations for one
LE using data from three European sites is illustrated in Figure 6.18. It can
be seen that the LEs calculated directly from measured data match well
with the simulated LEs. Similar results were found for all LEs and sites
considered.

6.5 ALLOWING FOR GROWTH IN VEHICLE
WEIGHTS AND FREQUENCIES

The quantity of road freight, measured in tonne-kilometers, is related to
economic activity. The European Commission (2018) show that, over a
twenty-year period, European growth in freight, with some deviations,
marginally exceeds the growth in Gross Domestic Product (GDP). In the

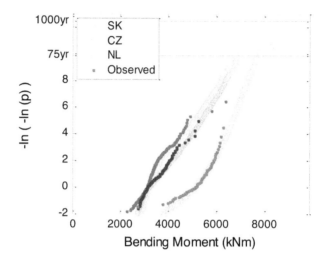

Figure 6.18 Bending moment at the center of a 30 m span simply supported bridge using observed data and simulations for three European sites (SK – Slovakia, CZ – Czech Republic, NL – the Netherlands).

United States, freight growth has tended to be less than GDP growth, perhaps because of the greater use of rail, but growth in road freight continues nonetheless. In simple terms, increases in the quantity of road freight result in greater numbers of heavier vehicles. There are very few studies in the literature addressing the influence of freight growth on bridge traffic loading. Indeed, the process of statistical extrapolation to find characteristic maximum values usually assumes that the underlying probability distribution does not change; that is, stationarity is assumed.

In this section a study of the influence of freight-traffic growth by OBrien et al. (2014) is reported. It is important to note that growth in road freight due to changes in vehicle types or regulations is not considered. Instead the work considers just one aspect of freight growth: an increase in the numbers of vehicles. Nevertheless, this increased frequency increases the probability that the maximum-per-day LE is the result of a heavier vehicle or combination of vehicles meeting or overtaking on the bridge.

6.5.1 Illustration of non-stationary extremes

As an illustration of the influence of a non-stationary process on the extremes, a simple example is considered using truck weights. Consider truck weights as Normally distributed with a mean of 50 t and a standard deviation of 5 t. Consider there to be 1000 trucks per day. Then the distribution of daily maximum weight is reasonably approximated by the Generalized Extreme Value (GEV) distribution. More details are given above and in Chapter 3, but recall that the GEV distribution is described by three parameters: location, scale and shape, and the shape parameter, ξ, determines if it is Weibull, Gumbel, or Fréchet. Next, consider the case where the number of trucks per day grows annually by 4.1%, which for 250 working days per year is an average daily growth of 0.016%. Thus, for each day, the size of the sample from which the maximum is taken, is a little bigger and the sample belongs to a slightly different GEV distribution. This example is illustrated in Figure 6.19: the daily maxima initially range from about 63 to 71 t but after 100 years, the range has shifted to about 68 to 75.

The parameter values from the non-stationary GEV distribution are shown in Figure 6.20. The location parameter can be seen to be increasing, the scale parameter is falling and the shape parameter is constant. This is reasonably typical parameter behavior, according to Coles (2001). A sample of daily maximum stationary and non-stationary data are illustrated on Gumbel paper in Figure 6.21. It is interesting to note how the non-stationary data has changes in slope like a mixture distribution and is shifted to the right. The difference in characteristic values between the growth and no-growth cases can be seen to be quite small – for example, for a 75-year return period, the difference in GVW is about 3.6%. It will be shown in the next section that the influence of non-stationarity on characteristic values can be much greater than the simple example here. Compounding this, in more realistic examples, the parent distribution may also be a mixture that includes very rare event types.

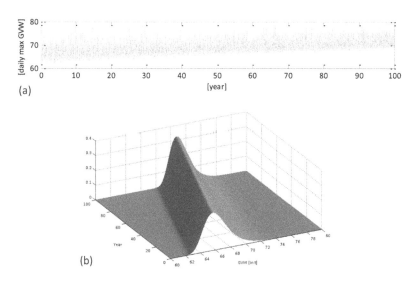

Figure 6.19 Non-stationary distribution: (a) maximum per day data over 100 years; (b) probability distribution changing through time (after OBrien et al. 2014).

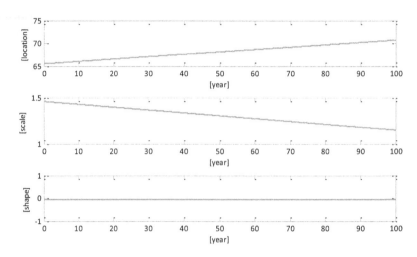

Figure 6.20 Changing parameter values of GEV distributions when vehicle numbers per day are growing (after OBrien et al. 2014).

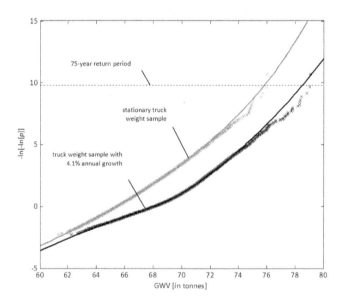

Figure 6.21 Probability paper plot of maximum-per-day weights where the number of vehicles is stationary (gray) and growing at 4.1% per year over 100 years (black).

6.5.2 Influence of growth on bridge traffic load effects

OBrien et al. (2014) report a series of simulations that explore the implications of growth in truck numbers on characteristic maximum LEs. They use a WIM database with over 600,000 trucks measured at a site in the

Netherlands in 2005. To explore the implications of growth in vehicle numbers, they consider the relationship between vehicle flow rate and inter-vehicle gaps. Statistical distributions are fitted to the measured gap data for 20 different flows. These in turn are used to generate typical inter-vehicle gaps as the flows change due to traffic volume growth. Growth in all types of vehicle is assumed over a 75-year design life for two-lane same-direction traffic. Block maximum LEs are calculated using blocks of 25 working days.

Mid-span moment is calculated in simply supported bridges assuming daily growth equivalent to annual growth rates of 0%, 1%, 2%, and 3%. Following a sensitivity study, growth is assumed to cause the location and scale parameters of the GEV distributions to vary linearly but not to affect the shape parameter. The results for four bridge spans are presented in Figure 6.22. For the longer spans, a change in the slope of the curves can be seen at a Standard Extremal Variate of around four, corresponding to a return period of about five years. This is shown to be the result of the block maximum LE changing from being predominantly due to crane-type vehicles to being predominantly due to low loaders. The effect of growth is to shift all curves to the right, and this is the predominant effect on the 75-year characteristic maximum value which, for the 30 m bridge, has increased by about 10% due to 3% annual growth. Where a change in slope occurs, the change tends to happen earlier in the growth scenario, but this does not significantly affect the characteristic value.

6.5.3 Recommendations for considering traffic growth

As has been seen, although intricate, the inclusion of simple forms of growth in both traffic models and hence the Extreme Value statistical framework,

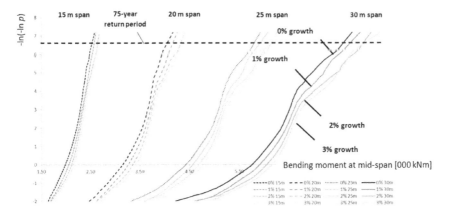

Figure 6.22 Average results from ten simulations of mid-span bending moment in bridges with four different spans and four different growth rates over a 75-year lifetime.

is reasonably straightforward (OBrien et al. 2014, Coles 2001). However, to be more realistic, models should also allow for changes in the relative frequencies of different truck configurations. For example, there is anecdotal evidence that 6-axle trucks are becoming increasingly common in Europe. The information required to complete such models, including knowledge of future regulatory changes, is clearly very difficult to predict, especially for an entire bridge design life.

For the assessment of existing bridges, the simplest approach is to base the analysis on measured site data, neglect traffic growth and monitor the situation. For example, the assessment could be assumed to be valid for, perhaps five years. During this period, traffic data would be used to confirm that the assessment remains valid. If the data indicates traffic growth, the assessment should be repeated. For the development of notional load models for new bridge design, allowances for traffic growth can be made through: (1) sensitivity studies and hypothesized traffic characteristics and vehicle populations, or (2) using the maximum possible truck volume and likely future freight density, as was done in Australia. Other approaches may also be possible.

6.6 TRAFFIC LOADING ON SECONDARY ROAD BRIDGES

WIM data is most often collected on primary routes and relatively little is known about traffic loading on bridges in secondary road networks. In general, there are far fewer extreme vehicles on secondary roads so the characteristic maximum LEs will typically be much less than for highways (noting that there may be exceptions for local 'last mile' roads near heavy industry). The challenge is to quantify the probabilities and estimate these characteristic maximum values. Section 6.4 addresses the issue of standard and non-standard vehicles. In this section, a more general 'extreme vehicle' is considered which may be a standard routine permit vehicle or a non-standard (special permit) vehicle – see Figure 6.11.

6.6.1 Sources of extreme vehicles

Central Place Theory was developed by Christaller (King 1984) to explain the location of cities and towns within a region, according to their function. It is used here to review the source/destination of heavy vehicles. The concept is that cities develop naturally to provide services to the surrounding region and the types of business in a city or town are influenced by the size of its population. The size of these market areas varies across different industries. For example, all towns require a grocery store whereas only larger towns or cities have more specialized functions such as a university.

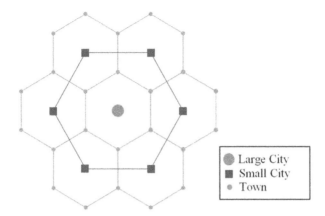

Figure 6.23 Hexagonal pattern for locating cities and towns.

The theory generates a hierarchy of central places, with the range of goods available increasing according to the size of the city/town. The theory assumes an infinite homogeneous plain that can be settled uniformly, without any constraining geographical features. Then each city/town within the central places hierarchy will be equidistant from other cities/towns of similar size. Christaller suggests that the resulting settlement pattern will be a hexagonal one – see Figure 6.23.

6.6.1.1 Standard vehicles

From a transportation perspective, each city/town is assumed to import goods from a larger city/town and to export to a smaller city/town (O'Sullivan 2009), usually using standard vehicles. For the most part, these vehicles carry consumer goods such as food, furniture, clothes, and electrical products between cities/towns. These generally have two or three axles on a rigid base or a 5- or 6-axle articulated configuration. Using Central Place Theory, it is assumed that these trucks are found mainly on the roads connecting larger cities with smaller ones and larger towns with smaller ones. Their number can be reasonably assumed to be proportional to city/town sizes.

6.6.1.2 Non-standard vehicles

Non-standard vehicles make up a very small proportion of the overall truck population (e.g. 1% to 2%) but they play a key role in critical bridge loading events for short to medium span bridges. Leahy (2014) finds that permit trucks are almost entirely related to the construction industry. From over 390 photographs of permit vehicles at a WIM site in the Netherlands, he identifies 328 as construction-related, 56 as inconclusive, and 6 as non-construction related. Construction is associated with areas which are experiencing economic growth. As the wealth of an urban or city region grows, new domestic and commercial buildings need to be constructed (O'Sullivan 2009) as well as new infrastructure such as bridges, roads, and tunnels. It is these construction projects that are the main destinations for permit trucks. Leahy (2014) suggests that, unlike standard freight movement, construction traffic is then not directly proportional to the size of nearby cities, but rather proportional to both the size of the city/town and its economic growth. Of course, permit vehicles may also be travelling to other locations dispersed throughout a country or region. For example, one-off developments such as wind turbines, power plants, and tunnels/bridges are often located away from large towns and cities.

6.6.2 Bayesian Updating and the 'megasite' concept

Bayesian Updating (Ang & Tang 2007), coming from Bayes Theorem, is a statistical method of combining prior knowledge with new evidence. In this case, the prior knowledge comes from a large database of WIM records from 19 American WIM sites, which Leahy (2014) refers to as the 'megasite.' The megasite provides baseline information on the traffic of the country or region, including the relative percentages of different vehicle types. The new evidence is the limited quantity of WIM data from the site of interest. For secondary roads in particular, there is generally a limited quantity of data, due to the lower volumes of traffic on these roads. While the composition of the traffic on primary and secondary routes may be different, even for the same volume, it is conservative to assume the same percentage of extreme vehicles on all types of road.

6.6.2.1 Bayesian Updating and Kernel Density Estimation

Considering the discussion in Section 6.2, the weekly maximum GVW across the megasite is taken as a (bounded) Weibull distribution. The parameters of this distribution are determined from all the WIM data across the megasite. This distribution is the prior. When a new set of WIM measurements are taken at the site of interest, these data are used to update the parameters of the Weibull distribution using Baye's Theorem. This 'posterior' distribution is a weighted mean of the data at the megasite and the site of interest.

For sparse data, Kernel Density Estimation (KDE) (Silverman 1986) can be used to approximate a continuous probability distribution to the prior data. In KDE, each data point can be replaced with a local Normal distribution centered on the data with a bandwidth (standard deviation). The resulting probability densities are added and normalized. If the bandwidth is narrow it matches the data very closely, sometimes over-fitting to a set of measurements. If, on the other hand, the bandwidth is large, the data may be oversmoothed and the detail of the trends in the data may be lost. Silverman (1986) recommends a trial and error approach to bandwidth selection.

6.6.2.2 The megasite data

The data used in this work is from the United States Federal Highway Administration's Long-Term Pavement Performance (LTPP) Program. The database has 38,430 site-days (105 site-years) of data and contains 81.6 million truck records. Leahy (2014) filters this data, as described in Section 6.4, to remove apparent non-standard vehicles. To give equal weighting to each site, four years of WIM data (2008-2011), is used from each site for the purpose of estimating the prior distributions. To estimate the distribution of the parameter vector, each year of data from each site is assessed separately, giving a total of 76 site-years – see Figure 6.24. As is clear, the properties of these 76 distributions vary greatly, as do the resulting characteristic maximum values.

The combinations of parameter values for the 76 site-years of data are shown as points in 3-dimensional space in Figure 6.25. To illustrate the permutations, values of σ are plotted against μ, for different ranges of the shape factor, ξ. The most frequent combination occurs around ($\mu = 60$, $\sigma = 12$, ξ: $0 - 0.2$) but other regions also feature frequently such as around ($\mu = 90$, $\sigma = 8$, ξ: $0.2 - 0.4$). The Maryland site stands out from the others, occurring around ($\mu = 45$, $\sigma = 5$, ξ: $0 - 0.2$).

Leahy (2014) uses two Kernel Density bandwidths to test the sensitivity of the results to the selected bandwidth. Bandwidth A ($BW_{\mu}^{B} = 3, BW_{\sigma}^{B} = 1, BW_{\xi}^{B} = 0.04$) fits the histograms of parameter values quite closely, perhaps too much so – see Figure 6.26. Bandwidth B

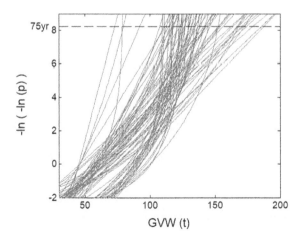

Figure 6.24 Weibull distributions fitted to each of the 76 site-years of prior data.

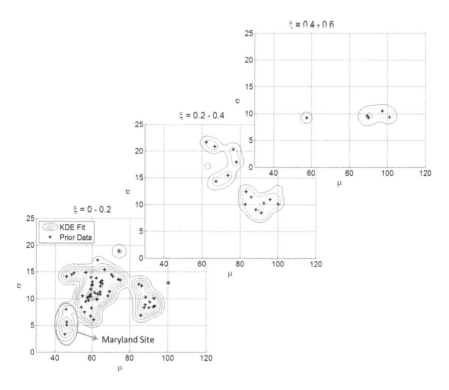

Figure 6.25 Prior data used for different ranges of shape parameter, ξ. Contours indicate the Kernel Density estimated probabilities of each parameter set (Bandwidth A: $BW_{\mu}^{A} = 3, BW_{\sigma}^{A} = 1, BW_{\xi}^{A} = 0.04$).

$(BW_{\mu}^{B} = 6, BW_{\sigma}^{B} = 2, BW_{\xi}^{B} = 0.08)$, in which the width of the Kernels (standard deviations of each small Normal distribution) have been doubled, smooths out much of the detail in the data and fits the more general trend.

6.6.2.3 Updating the characteristic maximum GVW for the Maryland site

The WIM data at the Maryland WIM site (US-15 North), analyzed by Leahy (2014), is chosen here as an example of a 'secondary' road. While it is designated as a highway, it is not an interstate and the WIM data (Figure 6.25) indicates significantly different statistical features from the other sites in the megasite database, many of which are interstates. The Maryland site has an average of 1030 trucks per weekday and the mean maximum GVW per week is 51 tonnes, the least for all sites in the database. Nearly seven years of WIM data is available for the site and all seven years of weekly maximum values are used in a conventional (non-Bayesian) fit to a Weibull distribution to determine a benchmark 75-year characteristic maximum

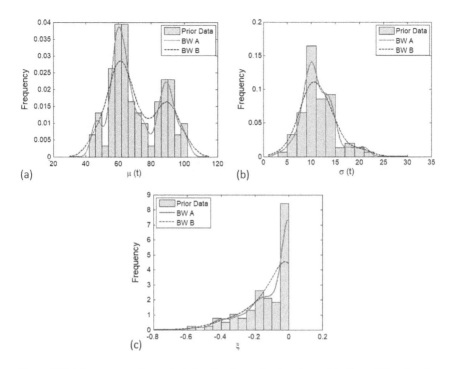

Figure 6.26 Histograms of marginal distributions of prior data with Kernel Density distribution for Bandwidths (BW) A and B (after Leahy 2014). (a) Location parameter; (b) Scale parameter; (c) Shape parameter.

value. This benchmark result is deemed to be accurate and is used to assess the accuracy of other approaches. The more mainstream Colorado site (I-76 East) is also considered. The goal of this study was to determine if good results could be obtained from an analysis of a much smaller database.

To simulate the condition of having a limited database, just 25 weekly maximum values, selected at random, are deemed to be available. The megasite, excluding Maryland, is used to enhance the prediction of 75-year characteristic maximum GVW obtained from these 25 values. This process is repeated 100 times and the accuracy, relative to the benchmark result, is calculated. Conventional (non-Bayesian) fits to each set of 25 values are also carried out to facilitate a comparison.

Results for the two sites are presented in Figure 6.27. For the Maryland site (Figure 6.27(a)), the accuracy of results is similar for the Bayesian and non-Bayesian approaches when the larger bandwidths are chosen (Bayesian B). The mean accuracy of both approaches is then below 2%, with the standard deviation being less for the Bayesian approach (10.4% versus 14.5% for Weibull). The results are sensitive to the bandwidths – the mean error for the Maryland site increases to +14.6% (conservative) when Bandwidth A is

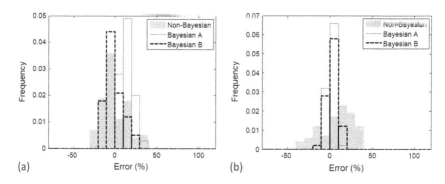

Figure 6.27 Histogram of the errors (relative to benchmark) in estimating 75-year GVW values using non-Bayesian and Bayesian (with two bandwidths) methods: (a) Maryland; (b) Colorado.

chosen. For the more typical WIM site at Colorado, the Bayesian approach significantly improves the accuracy and is insensitive to bandwidth, reducing the standard deviation of the error from 19.3% to about 5%. This may be because this site, being heavily trafficked, has parameter values already closer to the megasite.

It is difficult to draw firm conclusions from this single study. Nevertheless, the concept of adopting an approach that weights the results of a limited quantity of data with the trends established from a more reliable megasite, could be a useful way of reducing data collection costs. In lightly trafficked sites in particular, less data tends to be available and there is a risk that extreme vehicles are insufficiently represented in the data. The variability in the results, both for conventional and Bayesian approaches, is of concern and highlights the importance of having larger databases. In such circumstances, conservatism is advised.

6.7 DISCUSSION

The accuracy of characteristic maximum load effect calculations has increased significantly in recent decades and results are now more reliable and consistent than ever before. This is aided by the ready availability of very large WIM databases and great increases in computing power. Despite the size of some WIM databases, it is still not generally practical to calculate characteristic maximum values without some form of modelling or tail fitting. Much of this chapter involves a review of methods of extrapolation – POT, GEV, etc. It is recommended that Fréchet (unbounded) behavior should be prevented, i.e. the shape parameter, ξ, should be limited. Apart from that, the method of extrapolation does not greatly affect the accuracy of the results and many alternative approaches are acceptable. A more significant issue is how to identify the tail of the distribution. There is no simple answer to this except that (a) the tail should be beyond the last change in slope and (b) there should be enough data in the tail to represent a significant trend. Fortunately, big extrapolations are no longer necessary if a long-run simulation approach is adopted, something that is relatively easy to implement (Chapter 3). This addresses both the issue of which distribution to select and where the tail starts. It also allows typical extreme loading scenarios to be generated, providing insights into the nature of these loading events.

Issues other than tail fitting are critical for the calculation of characteristic maximum LEs. A great many studies do the calculations using all vehicles in the WIM database. This neglects the probability that at least some of these vehicles may be non-standard and should not be included in a calculation for normal loading. This issue can be effectively addressed by using axle configurations to filter apparent non-standard vehicles from the database.

Bridges with multiple same-direction lanes do not receive the attention they deserve as correlations between vehicle weights mean that lanes cannot be treated independently. Scenario modeling is an effective means of addressing this issue. Traffic growth is another phenomenon that is well known but receives little attention. One study is reported here that investigates the implications of increased vehicle frequencies, something that does not seem to greatly influence the results. However, many unanswered questions remain about other types of growth, such as shifts of freight from one vehicle type to another.

References

[1] http://arches.fehrl.org/index.php?m=7&id_directory=1611

23 U.S. Code § 127 (1974), 'Vehicle weight limitations - Interstate system', *TITLE 23 -HIGHWAYS*, US Code, 105–111.

AASHO (1928), *Conference specifications for steel highway bridges, compiled by a Conference Committee composed of State highway officials and the American Railway Engineering Association*, American Association of State Highway Officials, Washington, DC.

AASHO (1949), *Standard specifications for highway bridges adopted by the American Association of State Highway Officials*, 5th ed., American Association of State Highway Officials, Washington, DC.

AASHTO (1997), *Specification for highway bridges*, American Association of State Highway and Transportation Officials, Washington, DC.

AASHTO (2010), *AASHTO LRFD bridge design specifications*, 5th ed., American Association of State Highway and Transportation Officials, Washington, DC.

AASHTO (2011), *The manual for bridge evaluation*, 2nd ed., American Association of State Highway and Transportation Officials, Washington, DC.

AASHTO (2012), *AASHTO LRFD bridge design specifications*, 6th ed., American Association of State Highway and Transportation Officials, Washington, DC.

AASHTO (2014), *AASHTO LRFD bridge design specifications*, 7th ed., American Association of State Highway and Transportation Officials, Washington, DC.

AASHTO (2018), *Manual for bridge evaluation*, 3rd ed., American Association of State Highway and Transportation Officials, Washington, DC.

Abdel-Rohman, M., Al-Duaij, J. (1996), 'Dynamic response of hinged-hinged single span bridges with uneven deck', *Computers & Structures*, 59(2), 291–299.

Abramson, I.S. (1982), 'On bandwidth variation in kernel estimates-a square root law', *The Annals of Statistics*, 10(4), 1217–1223.

Agarwal, A.C., Wolkowicz, M. (1976), *Interim report on 1975 commercial vehicle survey*, Research and Development Division, Ministry of Transportation and Communications, Downsview, ON, Canada.

Akerman, I., Jonsson, R. (2007), *European modular system for road freight transport – Experiences and possibilities*, TFK – Transport Research Institute, Stockholm.

Ali, H., Nowak, A.S., Stallings, J.M., Chmielewski, J., Stawska, S., Haddadi, F. (2020), *Impact of Heavy Trucks and Permitted Overweight Loads on Highways and Bridges*, Report submitted to Florida Department of Transportation, May.

Allen, T.M., Nowak, A.S., Bathurst, R.J. (2005), *Calibration to determine load and resistance factors for geotechnical and structural design*, Transportation Research Board. https://www.nap.edu/catalog/21978/calibration-to-determine-load-and-resistance-factors-for-geotechnical-and-structural-design (accessed 15 November 2016).

Al-Qadi, I., Wang, H., Ouyang, Y., Grimmelsman, K., Purdy, J. (2016), *LTBP program's literature review on weigh-in-motion systems*, No. FHWA-HRT-16-024, Research, Development, and Technology, Turner-Fairbank Highway Research Center, McLean, VA.

Ang, A.H.-S., Tang, W.H. (2007), *Probability concepts in engineering : Emphasis on applications in civil & environmental engineering*, 2nd ed., Wiley, New York.

Anitori, G., Casas, J.R., Ghosn, M. (2017), 'WIM-based live-load model for advanced analysis of simply supported short- and medium-span highway bridges', *Journal of Bridge Engineering*, 22, 04017062. doi: 10.1061/(ASCE) BE.1943-5592.0001081.

ASCE (1981), 'Committee on loads and forces on bridges – Recommended design loads for bridges', *Journal of the Structural Division*, ASCE, 107, 1161–1213.

ASCE (2017), *America's infrastructure report card 2017, GPA: D+*, American Society of Civil Engineers, https://www.infrastructurereportcard.org/ (accessed 11 June 2020).

ASTM (2009), *E1318 - 09 - Standard specification for highway weigh-in-motion (WIM) systems with user requirements and test methods*, ASTM International, West Conshohocken, PA.

ASTM (2015), *Standard practice for computing international roughness index of roads from longitudinal profile measurements*, ASTM E1926 - 08(2015), American Society for Testing and Materials.

Atkins Highways and Transportation (2005), *Background to the UK national annexes to EN1990: Basis of structural design - Annex A2: Application for bridges EN1991-2: Traffic loads on bridges*.

Austroads (1996), *Austroads bridge design code*, Austroads, Sydney.

Austroads (2003), *Dynamic interaction of vehicles and bridges*, Publication No. AP-T23–03, Austroads, Sydney.

Babu, A.R., Iatsko, O., Nowak, A.S., Stallings, J.M. (2019), 'Improving quality of WIM traffic data', *98th Annual Meeting of the Transportation Research Board*, Washington, DC, Paper No. 19-01057.

Babu, A.R., Iatsko, O., Stallings, J.M., Nowak, A.S. (2019), *Application of WIM and permit data*, Report 930-947, ALDOT, Montgomery, AL.

Bacharz, M., Chmielewski, J., Stawska, S., Bacharz, K., Nowak, A. (2020), *Comparative analysis of vehicle weight measurement techniques – Evaluation of SiWIM system accuracy*, Research Report, Dept. of Civil Engineering, Auburn University, USA.

Bailey, S.F. (1996), *Basic principles and load models for the structural safety evaluation of existing bridges*, Thesis No 1467, École Polythechnique Fédéral de Lausanne.

Bailey, S.F., Bez, R. (1999), 'Site specific probability distribution of extreme traffic action effects', *Probabilistic Engineering Mechanics*, 14(1), 19–26. doi: 10.1016/S0266-8920(98)00013-7.

Bali, T.G. (2003), 'The generalized extreme value distribution', *Economics Letters*, 79(3), 423–427.

Bando, M., Hasebe, K., Nakayama, A., Shibata, A., Sugiyama, Y. (1995), 'Dynamical model of traffic congestion and numerical simulation' *Physical Review E*, 51,1035–1042.

Barlow, W.H. (1867), 'Decription of the Clifton Suspension Bridge (including plate)'. *Minutes of the Proceedings of the Institution of Civil Engineers*, 26, 243–257.

Barlow, W.H., Claxton Hemans, G.W., Brereton, R.P., Fowler, J., Royal, A., Radford, G.K., Law, H., Gregory, C.H., Richardson, R., Bergue, D., Cochrane Longridge, J.A., Vignoles Reilly, C., Feuvre, L., Hawkshaw Barlow, P., Cowper, E.A., Shields, F.W., Porter, J.H., Bidder, G.P., Lane, C.B., Bramwell, F.J., Knipple, W.R., Johnson, T.M., Williams, P., Fowler, T., Young, E.W., Phipps, G.H., Heppel, J.M., Whiting (1867), 'Discussion: Suspension bridges of great span', *Minutes of the Proceedings of the Institution of Civil Engineers*, 26, 265–309.

Barndorff-Nielsen, O. (1983), 'On a formula for the distribution of the maximum likelihood estimator', *Biometrika*, 70, 343–365.

Basson, S.E., Lenner, R. (2019), 'Reliability verification of bridges designed according to TMH-7', *Advances in Engineering Materials, Structures and Systems: Innovations, Mechanics and Applications - Proceedings of the 7th International Conference on Structural Engineering, Mechanics and Computation*, 1865–1870. doi: 10.1201/9780429426506-322.

Beck, J.L., Au, S.-K. (2002), 'Bayesian updating of structural models and reliability using Markov Chain Monte Carlo simulation', *Journal of Engineering Mechanics*, 128(4), 380–391.

Bermudez, P.D.Z., Kotz, S. (2010), 'Parameter estimation of the generalized pareto distribution – Part ii', *Journal of Statistical Planning and Inference*, 140(6), 1374–1388.

Bhattacharya, S.K. (1967), 'Bayesian approach to life testing and reliability estimation', *Journal of the American Statistical Association*, 62(317), 48–62.

Botev, Z.I., Grotowski, J.F., Kroese, D.P. (2010), 'Kernel density estimation via diffusion', *The Annals of Statistics*, 38(5), 2916–2957.

Box, G.E.P., Cox, D.R. (1964), 'An analysis of transformations', *Journal of the Royal Statistical Society, Series B (Methodological)*, 26(2), 211–252.

Brady, S.P., OBrien, E.J. (2006), 'The effect of vehicle velocity on the dynamic amplification of two vehicles crossing a simply supported bridge', *Journal of Bridge Engineering*, ASCE, 11(2), 250–256.

Brady, S.P., OBrien, E.J., Žnidarič, A. (2006), 'The effect of vehicle velocity on the dynamic amplification of a vehicle crossing a simply supported bridge', *Journal of Bridge Engineering*, ASCE, 11(2), 241–249.

Brockfeld, E., Kühne, R.D., Wagner, P. (2004), 'Calibration and validation of microscopic traffic flow models', *Transportation Research Record: Journal of the Transportation Research Board*, 1876(1), 62–70.

Broquet, C., Bailey, S.F., Fafard, M., Brühwiler, E. (2004), 'Dynamic behavior of deck slabs of concrete road bridges', *Journal of Bridge Engineering*, 9(2), 137–146.

Brühwiler, E., Herwig, A. (2008), 'Consideration of dynamic traffic action effects on existing bridges at ultimate limit state', *Proc. of IABMAS'08: 4th Int. Conf. on Bridge Maintenance, Safety and Management*, Eds. H.M. Koh and D.M. Frangopol, Taylor & Francis, London, 3675–3682.

Bruls, A., Croce, P., Sanpaolesi, L. (1996b), 'Calibration of road load models for road bridges', *ENV1991 - Part 3: Traffic loads on bridges: Calibration of road load models for road bridges*, IABSE Report, 439–453.

Bruls, A., Croce, P., Sanpaolesi, L., Sedlacek, G. (1996a), 'ENV1991 – Part 3: Traffic loads on bridges; calibration of load models for road bridges', *Proceedings of IABSE Colloquium*, Delft, The Netherlands, 439–453.

BSI (1954), *BS 153: Part 3A: 1954 girder bridges - Part 3 - Loads and stress - Section A: Loads*, British Standards Institution, London.

BSI (1978), *BS 5400 steel, concrete, and composite bridges. Part 2 specification for loads*, British Standards Institution, London.

BSI (2006), *BS 5400, 2006, part 2: Steel, concrete and composite bridges. Specification for loads*, British Standards Institution, London.

Buckland, P.G. (1981), 'Recommended design loads for bridges (Committee on loads and forces on bridges of the committee on bridges of the structural division)', *Journal of the Structural Division*, ASCE, 1161–1213.

Buckland, P.G., McBryde, J.P., Navin, F.P.D., Zidek, J.V. (1978), 'Traffic loading of long span bridges', *Transportation Research Record No. 665, Bridge Engineering*, 2.

Buckland, P.G., McBryde, J.P., Zidek, J.V., Navin, F.P.D. (1980), 'Proposed vehicle loading of long-span bridges', *Journal of the Structural Division*, ASCE, 106, 915–932.

Buckland, P.G., Navin, F.P.D., Zidek, J.V. (1975), 'Bridge traffic loads - Are we overdesigning?', Roads and Transportation Association of Canada.

Butler, R.W. (1986), 'Predictive likelihood inference with applications', *Journal of the Royal Statistical Society, Series B (Methodological)*, 48(1), 1–38.

Calgaro, J., Sedlacek, G. (1992), 'EC 1: Traffic loads on road bridges', *Structural Eurocodes – International Association for Bridge and Structural Engineering (IABSE) Conference*, September 14–16, Davos, Switzerland, 81–87.

CALTRANS (2004), *Bridge design specifications*, California Department of Transportation, Sacramento, CA.

Cambridge Systematics (2007), *MAG internal truck travel survey and truck model development study*, Maricopa Association of Governments.

Cantero, D., González, A., OBrien, E.J. (2009), 'Maximum dynamic stress on bridges traversed by moving loads', *ICE Bridge Engineering*, 162(2), 75–85.

Cantero, D., González, A., OBrien, E.J. (2011), 'Comparison of bridge dynamic amplification due to articulated 5-axle trucks and large cranes', *The Baltic Journal of Road and Bridge Engineering*, 6(1), 39–47.

Cantieni, R. (1983), *Dynamic load tests on highway bridges in Switzerland –60 years experience of EMPA*, Report No. 211, EMPA, Swiss Federal Laboratories for Materials Testing and Research, Dübendorf, Switzerland.

Caprani, C.C. (2005), *Probabilistic analysis of highway bridge traffic loading*, PhD Dissertation, Dept. of Civil Engineering, University College Dublin, Ireland.

Caprani, C.C. (2010), 'Using microsimulation to estimate highway bridge traffic load', *Proceedings of 5th International Conference on Bridge Maintenance, Safety and Management*, Eds. R. Sause and D. Frangopol, IABMAS, CRC Press, Philadelphia, PA.

Caprani, C.C. (2012a), 'Calibration of a congestion load model for highway bridges using traffic microsimulation', *Structural Engineering International*, 22(3), 342–348. doi: 10.2749/101686612X13363869853455.

Caprani, C.C. (2012b), 'Bridge-to-vehicle communication for traffic load mitigation', *Proceedings of Bridge and Concrete Research in Ireland*, Eds. C. Caprani and A. O'Connor, Dublin Institute of Technology and Trinity College Dublin, Dublin, Ireland, 25–30.

Caprani, C.C. (2013), 'Lifetime highway bridge traffic load effect from a combination of traffic states allowing for dynamic amplification', *Journal of Bridge Engineering*, 18(9), 901–909.

Caprani, C.C. (2018), 'A problem ignored: Highway bridges and automated truck platoons', *Infrastructure Magazine*, 6, 14–17.

Caprani, C.C., Belay, A., O'Connor, A. (2003), 'Site-specific probabilistic load modelling for bridge reliability analysis', *Proceedings of the 3rd International Conference on Current and Future Trends in Bridge Design*, 341–348.

Caprani, C.C., De Maria, J. (2020), 'Long-span bridges: Analysis of trends using a global database', *Structure and Infrastructure Engineering*, 16(1), 219–231. doi: 10.1080/15732479.2019.1639773.

Caprani, C.C., Enright, B., Carey, C. (2012a), 'Lane changing control to reduce traffic load effect on long-span bridges', *International Conference on Bridge Maintenance, Safety and Management*, IABMAS, Lake Como, Italy, July.

Caprani, C.C., González, A., Rattigan, P.H., OBrien, E.J. (2012b), 'Assessment dynamic ratio for traffic loading on highway bridges', *Structure and Infrastructure Engineering*, 8(3), 295–304.

Caprani, C.C., OBrien, E.J. (2006), 'Statistical computation for extreme bridge traffic load effects', *Proceedings of the 8th International Conference on Computational Structures Technology*, Civil-Comp Press, Stirling, Scotland.

Caprani, C.C., OBrien, E.J. (2008), 'The governing form of traffic for highway bridge loading', *Proceedings of 4th Symposium on Bridge and Infrastructure Research in Ireland*, Eds. E. Cannon, R. West and P. Fanning, National University of Ireland, Galway, 53–60.

Caprani, C.C., OBrien, E.J. (2009), 'Estimating extreme highway bridge traffic load effects', *International Conference on Structural Safety and Reliability, ICOSSAR '09*, Eds. H. Furuta, D. Frangopol, and M. Shinozuka, CRC Press, Osaka, Japan.

Caprani, C.C., OBrien, E.J. (2010), 'The use of predictive likelihood to estimate the distribution of extreme bridge traffic load effect', *Structural Safety*, 32(2), 138–144.

Caprani, C.C., OBrien, E.J, Lipari, A. (2016), 'Long-span bridge traffic loading based on multi-lane traffic micro-simulation', *Engineering Structures*, 115, 207–219. doi: 10.1016/j.engstruct.2016.01.045.

Caprani, C.C., OBrien, E.J., McLachlan, G. (2008), 'Characteristic traffic load effects from a mixture of loading events on short to medium span bridges', *Structural Safety*, 30(5), 394–404.

Carey, C., Caprani, C.C., Enright, B. (2012), 'Reducing traffic loading on long-span bridges by means of lane-changing restrictions', *Proceedings of Bridge and Concrete Research in Ireland*, Eds. C. Caprani and A. O'Connor, Dublin Institute of Technology and Trinity College Dublin, Dublin, Ireland, 331–336.

Carey, C., Caprani, C.C., Enright, B. (2018), 'A pseudo-microsimulation approach for modelling congested traffic loading on long-span bridges', *Structure and Infrastructure Engineering*, 14(2), 163–176. doi: 10.1080/15732479.2017.1330893.

Carey, C., OBrien, E.J., Malekjafarian, A., Lydon, M., Taylor, S. (2017), 'Direct field measurement of the dynamic amplification in a bridge', *Mechanical Systems and Signal Processing*, 85, 601–609.

Castillo, E. (1988), *Extreme value theory in engineering*, Academic Press, London, Retrieved from http://www.sciencedirect.com/science/book/9780121634759.

Castillo, E., Solares, C., Gomez, P. (1996), 'Probabilities using tail simulated data', *International Journal of Approximate Reasoning*, 17, 163–189.

CDOT (2012), *Bridge design manual*, Colorado Department of Transportation, Denver, CO.

Cebon, D. (2000), *Handbook of vehicle-road interaction*, CRC Press, Boca Raton, FL.

CEN (2002), *Eurocode 1: Actions on structures, Part 1–1: General actions - Densities, self-weight, imposed load for building. European Standard EN 1991-1-1-1*, European Committee for Standardization, Brussels.

CEN (2003), *Eurocode 1: Actions on structures, Part 2: Traffic loads on bridges. European Standard EN 1991–2:2003*, European Committee for Standardization, Brussels.

Chan, T.H.T., O'Connor, C. (1990), 'Vehicle model for highway bridge impact', *Journal of Structural Engineering*, ASCE, 116(7), 1772–1793.

Chang, D., Lee, H. (1994), 'Impact factors for simple-span highway girder bridges', *Journal of Structural Engineering*, 120(3), 704–715.

Chen, C., Li, L., Hu, J., Geng, C. (2010), 'Calibration of MITSIM and IDM car-following model based on NGSIM trajectory datasets', *International Conference on Vehicular Electronics and Safety*, QingDao, China, 48–53.

Chen, S.R., Wu, J. (2011), 'Modelling stochastic live load for long-span bridge based on microscopic traffic flow simulation', *Computers & Structures*, 89(9–10), 813–824.

Chompooming, K., Yener, M. (1995), 'The influence of roadway surface irregularities and vehicle deceleration on bridge dynamics using the method of lines', *Journal of Sound and Vibration*, 183(4), 567–589.

Coles, S. (2001), *An introduction to statistical modeling of extreme values*, Springer, London.

Coles, S., Tawn, J. (1996), 'Modelling extremes: A bayesian approach', *Applied Statistics*, 45, 463–478.

Cooper, D.I. (1995), 'The determination of highway bridge design loading in the United Kingdom from traffic measurements', *Pre-Proceedings of the First European Conference on Weigh-in-Motion of Road Vehicles*, Eds. B. Jacob et al., E.T.H. Zürich, 413–421.

Cooper, D.I. (1997), 'Development of short span bridge-specific assessment live loading', *Safety of bridges*, Ed. P.C. Das, Thomas Telford, 64–89.

Cornell, C.A. (1968), 'Engineering seismic risk analysis', *Bulletin of the Seismological Society of America*, 58, 1583–1606.

COST 323 (2002), *Weigh-in-motion of road vehicles - Final report*, Eds. B. Jacob, E. OBrien, and S. Jehaes, LCPC, Paris.

COWI (1990), *Stoerbelt east bridge – Design basis*. BBD.

Cremona, C. (2001), 'Optimal extrapolation of traffic load effects', *Structural Safety*, 23, 31–46.

Cremona, C., Carracilli, J. (1998), 'Evaluation of extreme traffic loads effects in cable stayed and suspension bridges by use of WIM records', *2nd European Conference on Weigh-in-Motion of Road Vehicles*, Eds. E.J. OBrien and B. Jacob, 243–251.

Crespo-Minguillón, C., Casas, J.R. (1997), 'A comprehensive traffic load model for bridge safety checking', *Structural Safety*, 19(4), 339–359. doi: 10.1016/S0167-4730(97)00016-7.

Croce, P., Salvatore, W. (1998), 'Stochastic modelling of traffic loads for long-span bridges', *Long-Span and High-Rise Structures*, IABSE Symposium, Kobe, 427–434.

Croce, P., Salvatore, W. (2001), 'Stochastic model for multilane traffic effects on bridges', *Journal of Bridge Engineering*, 6(2), 136–143.

CSA (2006), *Canadian highway bridge design code*, Canadian Standards Association, Mississauga, ON.

Csagoly, P.F., Dorton, R.A. (1978), 'The development of the Ontario highway bridge design code', *1st Bridge Engineering Conference*, St Louis, Transportation Research Board, 1–12.

CSRA (1991), *Technical manual for highways 7: Part 2 traffic loading*. Code of Practice for the Design of Highway Bridges and Culverts in South Africa, Part 2: Traffic Loading (1991), Committee of State Road Authorities, Department of Transport, Pretoria, South Africa.

Dawe, P. (2003), *Traffic loading on highway bridges, research perspectives*, Thomas Telford, UK. doi: 10.1680/tlohb.32415.

De Angelis, D., Young, G.A. (1992), 'Smoothing the bootstrap', *International Statistical Review/Revue Internationale de Statistique*, 60(1), 45–56.

De Boer, P.T., Kroese, D.P., Mannor, S., Rubinstein, R.Y. (2005), 'A tutorial on the cross-entropy method', *Annals of Operations Research*, 134(1), 19–67.

De Ceuster, G., Breemersch, T., Herbruggen, B., Van Verweij, K., Davydenko, I. (2008), *Effects of adapting the rules on weights and dimensions of heavy commercial vehicles as established within Directive 96/53/EC*, European Commission, Brussels.

de Wet, D. (2010), 'WIM calibration and data quality management', *Journal of the South African Institution of Civil Engineering*, 52(2), 70–76.

Deng, L., Yu, Y., Zou, Q., Cai, C.S. (2015), 'State-of-the-art review of dynamic impact factors of highway bridges', *Journal of Bridge Engineering*, ASCE, 20(5). doi: 10.1061/(ASCE)BE.1943-5592.0000672.

Department of Transport (1992), *BD50/92 assessment and strengthening of highway structures, stage 3 – Long span bridges*, The Highways Agency, UK.

Dissanayake, P.B.R., Karunananda, P.A.K. (2008), 'Reliability index for structural health monitoring of aging bridges', *Structural Health Monitoring*, 7(2), 175–183.

Ditlevsen, O., Madsen, H. (1994), 'Stochastic vehicle-queue-load model for large bridges', *Journal of Engineering Mechanics*, 120, 1829–1847.

DIVINE (1997), *Dynamic interaction of heavy vehicles with roads and bridges*, Technical report, OECD, DIVINE Concluding Conference, Ottawa, ON.

DIVINE (1998), *Dynamic interaction between vehicles and infrastructure experiment*, OECD, Retrieved from www.oecd.org.

Dowling, J., OBrien, E.J., González, A. (2012), 'Adaptation of cross entropy optimisation to a dynamic bridge WIM calibration problem', *Engineering Structures*, 44, 13–22.

DRD (2004), *Reliability-based classification of the load carrying capacity of existing bridges. Guideline document*, Report 291, Danish Roads Directorate, Copenhagen.

Einmahl, M. (2008), 'Records in athletics through extreme-value theory', *Journal of the American Statistical Association*, 103(484), 1382–1391.

Elkins, G.E., Thompson, T., Ostrom, B., Visintine, B. (2018), *Long-term pavement performance information management system user guide*, U.S. Department of Transportation Federal Highway Administration, 277.

Elkins, L., Higgins, C. (2008), *Development of truck axle spectra from Oregon weigh-in-motion data for use in pavement design and analysis*, SPR 635, FHWA-OR-RD-08-06 Final Report, Oregon State University, Corvallis, OR.

Enright, B. (2010), *Simulation of traffic loading on highway bridges*, PhD Thesis, University College Dublin.

Enright, B., Carey, C., Caprani, C.C. (2013), 'Microsimulation evaluation of eurocode load model for American long-span bridges', *Journal of Bridge Engineering*, 18(12), 1252–1260.

Enright, B., Carey, C., Caprani, C.C., OBrien, E.J. (2012a), 'The effect of lane changing on long-span highway bridge traffic loading', Eds. F. Biondini and D.M. Frangopol, *Bridge maintenance, safety, management, resilience & sustainability*, Sixth International IABMAS Conference, Lake Maggiore, Stresa, Taylor & Francis, Italy.

Enright, B., Leahy, C., OBrien, E.J. Keenahan, J. (2012b), 'Portable bridge WIM data collection strategy for secondary roads', *Sixth International Conference on Weigh-In-Motion*, Eds. B. Jacob, A.-M. McDonnell, F. Schmidt, and C. Wiley, Dallas, TX, 402–409.

Enright, B., OBrien, E.J. (2010), 'Management strategies for special permit vehicles for bridge loading', *Transport Research Arena Europe 2010*, 7–10th June, Brussels, Belgium.

Enright, B., OBrien, E.J. (2011), *Cleaning weigh-in-motion data: Techniques and recommendations*, University College Dublin & Dublin Institute of Technology, Retrieved from http://iswim.free.fr/doc/wim_data_cleaning_ie .pdf.

Enright, B., OBrien, E.J. (2013), 'Monte Carlo simulation of extreme traffic loading on short and medium span bridges', *Structure and Infrastructure Engineering*, 9(12), 1267–1282. doi: 10.1080/15732479.2012.688753.

Enright, B., OBrien, E.J., Leahy, C. (2016), 'Identifying and modelling permit trucks for bridge loading', *Bridge Engineering, Proceedings of the Institution of Civil Engineers*, 169(4), 235–244. doi: 10.1680/bren.14.00031.

Enright, M.P., Frangopol, D.M. (1999), 'Condition prediction of deteriorating concrete bridges using Bayesian updating', *Journal of Structural Engineering*, 125(10), 1118–1125.

European Commission (2018), *EU transport in figures – Statistical pocketbook 2018*, Publications Office of the European Union, Luxembourg.

Felici, M., Lucarini, V., Speranza, A., Vitolo, R. (2007), 'Extreme value statistics of the total energy in an intermediate-complexity model of the midlatitude atmospheric jet. Part II: Trend detection and assessment', *Journal of the Atmospheric Sciences*, 64, 2159–2175.

FHWA (1995), *Traffic monitoring guide*, 3rd ed., U.S. Department of Transportation Federal Highway Administration.

FHWA (2001), *Traffic monitoring guide*, Federal Highway Administration, U.S. Department of Transportation, Office of Highway Policy Information.

FHWA (2016), *Traffic monitoring guide*, Office of Highway Policy Information, Federal Highway Administration, USA.

FHWA (2017), *FHWA's highway performance monitoring system (HPMS) division review guidelines*, Office of Highway Policy Information, Federal Highway Administration, USA.

FHWA (2018), *Weigh-in-motion data, 2014–2017*, US states: Alabama, California, Washington, DC, Florida, Montana, Rhode Island, South Dakota (accessed 2018).

Fiorillo, G., Ghosn, M. (2014), 'Procedure for statistical categorization of overweight vehicles in a WIM database', *Journal of Transportation Engineering*, ASCE, 140, 04014011. doi: 10.1061/(ASCE)TE.1943-5436.0000655.

Fisher, J., Mertz, D., Zhong, A. (1983), *Steel bridge members under variable amplitude, long life fatigue loading*, Final Report Draft, May 83pp. Fritz Laboratory Reports.

Fisher, R.A., Tippett, L.H.C. (1928), 'Limiting forms of the frequency distribution of the largest or smallest member of a sample', *Mathematical Proceedings of the Cambridge Philosophical Society*, 24(02), 180–190.

Flint and Neill Partnership (1986), *Interim design standard: Long span bridge loading*, Transport and Road Research Laboratory, Crowthorne.

Flint, A.R., Jacob, B.A. (1996), 'Extreme traffic loads on road bridges and target values for their effects for code calibration', *Proceedings of IABSE Colloquium*. Delft, The Netherlands, IABSE-AIPC-IVBH, 469–478.

Frýba, L. (1996), *Dynamics of railway bridges*, Thomas Telford, London.

Fu, G., Asce, M., You, J. (2010), 'Extrapolation for future maximum load statistics', *Journal of Bridge Engineering*, (August), 16, 527–535. doi: 10.1061/(ASCE)BE.1943-5592.0000175.

Fu, G., Hag-Elsafi, O. (1995), 'Bridge evaluation for overloads including frequency of appearance', *Applications of statistics and probability*, Eds. J.L. Favre and A. Mébarki, Rotterdam, 687–692.

Fu, G., You, J. (2011), 'Truck load modeling and bridge code calibration', *Applications of statistics and probability in civil engineering*, Eds. M. Faber, J. Kohler, and K. Nishijima, Taylor & Francis Group, London, 406–413.

Fwa, T.F., Li, S. (1995), 'Estimation of lane distribution of truck traffic for pavement design', *Journal of Transportation Engineering*, 121, 241–248.

Getachew, A. (2003), *Traffic load effects on bridges*, Doctoral thesis, Royal Institute of Technology (KTH), Sweden.

Getachew, A., OBrien, E.J. (2007), 'Simplified site-specific traffic load models for bridge assessment', *Structure and Infrastructure Engineering*, 3(4), 303–311.

Ghosn, M., Moses, F. (1985), 'Markov renewal model for maximum bridge loading', *Journal of Engineering Mechanics*, ASCE, 111(9), 1093–1104.

Ghosn, M., Moses, F. (1986), 'Reliability calibration of bridge design code', *Journal of Structural Engineering*, 112(4), 745–763. doi: 10.1061/ (ASCE)0733-9445(1986)112:4(745).

Ghosn, M., Sivakumar, B., Miao, F. (2010), 'Calibration of load and resistance factor rating methodology in New York State', *Transportation Research Record: Journal of the Transportation Research Board*, (2200). https://trid.trb.org/View/1082986?ajax=1 (accessed 6 October 2017).

Ghosn, M., Xu, Q. (1989), 'Estimating bridge dynamics using the weigh-in-motion algorithm', *Transportation Research Record*, 1200, 7–14.

Gibbons, N., Fanning, P.J. (2012), 'Rationalising assessment approaches for masonry arch bridges', *Proceedings of the ICE-Bridge Engineering*, 165(3), 169–184.

Gilleland, E., Katz, R.W. (2011), 'New software to analyze how extremes change over time', *Eos*, 11 January, 92(2), 13–14.

Gindy, M., Nassif, H.H. (2006a), 'Multiple presence statistics for bridge live load based on weigh-in-motion data', Transportation Research Board 86th Annual Meeting, Washington, DC.

Gindy, M., Nassif, H.H. (2006b), 'Comparison of traffic load models based on simulation and measured data', *Joint International Conference on Computing and Decision Making in Civil Engineering*, Eds. H. Rivard, E. Miresco, and H. Melhem, Montréal, 2497–2506.

Goldberg, D.E. (1989), *Genetic algorithms in search, optimization, and machine learning*, Addison-Wesley Publishing, Reading, MA.

González, A., Cantero, D., OBrien, E.J. (2011), 'Dynamic increment for shear force due to heavy vehicles crossing a highway bridge', *Computers and Structures*, 89(23–24), December, 2261–2272.

González, A., Rattigan, P., OBrien, E. J., Caprani, C. (2008), 'Determination of bridge lifetime dynamic amplification factor using finite element analysis of critical loading scenarios', *Engineering Structures*, 30(9), 2330–2337.

González, A., Žnidarič, A., Casas, J.R., Enright, B., OBrien, E.J., Lavric, I., Kalin, J. (2009), *Recommendations on dynamic amplification allowance*, Deliverable D10 of European 6th Framework ARCHES project, Forum of European National Highway Research Laboratories, http://arches.fehrl.org/?m=7&id_directory=1617.

Grave, S.A. (2001), *Modelling of site-specific traffic loading on short to medium span bridges*, PhD thesis, Trinity College Dublin, Ireland.

Grave, S.A., OBrien, E.J., O'Connor, A. (2000), 'The determination of site-specific imposed traffic loadings on existing bridges', *The Fourth International Conference on Bridge Management*, Eds. M.J. Ryal, G.A.R. Parke, and J.E. Harding., Thomas Telford, London.

Grundy, P., Grundy, J., Khalaf, H., Casagrande, R. (2002), 'Accommodation of time dependent drift of WIM data', *Third International Conference on Weigh-in-Motion (ICWIM3)*, Eds. B. Jacob, B. McCall, and E.J. OBrien, Orlando, 287–294.

Gumbel, E. (1935), 'Les valeurs extrêmes des distributions statistiques', Annales de l'Institut Henri *Poincaré*, 5(2), 115–158.

Guo, D., Caprani, C.C. (2019), 'Traffic load patterning on long span bridges: A rational approach', *Structural Safety*, 77, 18–29. doi: 10.1016/j.strusafe.2018.11.003.

Hajializadeh, D., OBrien, E.J., Enright, B., Caprani, C.C., Sheils, E., Wilson, S.P. (2012), 'Probabilistic study of lifetime load effect distribution of bridges', *6th International ASRANet Conference for Integrating Structural Analysis, Risk and Reliability*, London, UK.

Hallenbeck, M., Weinblatt, H. (2004), *Equipment for collecting traffic load data*, NCHRP Report No. 509, Transportation Research Board, Washington, DC. doi: 10.17226/13717.

Han, W., Wu, J., Cai, C.S., Chen, S. (2014), Characteristics and dynamic impact of overloaded extra heavy trucks on typical highway bridges, *Journal of Bridge Engineering*, 20(2), p. 05014011.

Hanswille, G., Sedlacek, G. (2007), *Background report traffic loads on road bridges basis of the load models in EN 1991–2 and DIN - Report 101*, Deutsches Institut Fur Normung E.V., Berlin, Germany.

Harman, D., Davenport, A. (1976), *The formulation of vehicular loading for design of highway bridges in Ontario*, Ontario Joint Transportation and Communications Research Program, Ontario.

Harman, D.J., Davenport, A.G. (1979), 'A statistical approach to traffic loading on highway bridges', *Canadian Journal of Civil Engineering*, 6, 494–513.

Harman, D.J., Davenport, A.G., Wong, W.S.S. (1984), 'Traffic loads on medium and long span bridges', *Canadian Journal of Civil Engineering*, 11, 556–573.

Harris, N.K., OBrien, E.J., González, A. (2007), 'Reduction of bridge dynamic amplification through adjustment of vehicle suspension damping', *Journal of Sound and Vibration*, 302, 471–485.

Hayrapetova, A.A. (2006), *Micro-simulation modelling of traffic loading on long span bridges*, PhD thesis, University College Dublin, Ireland.

Heitner, B., OBrien, E.J., Schoefs, F., Yalamas, T., Décatoire, R., Leahy, C. (2016), 'Evaluation of bridge safety based on weigh-in-motion data', *Civil Engineering Research in Ireland*, 2016, 3–8.

Helbing, D., Treiber, M., Kesting, A., Schönhof, M. (2009), 'Theoretical vs. empirical classification and prediction of congested traffic states', *The European Physical Journal B*, 69, 583–598. doi: 10.1140/epjb/e2009-00140-5.

Henderson, W. (1954), 'British highway bridge loading', *Proceedings of the Institution of Civil Engineers*, 3, 325–350.

Henderson, W., Burt, M.E., Goodearl, K.A. (1973), *Bridge loading, steel box girder bridges*. Institution of Civil Engineers, London.

Hendy, C., Mundell, C., Bishop, D. (2015), 'Management of the severn bridge suspension bridge, *International Conference on Multi-Span Large Bridges*, Eds. A. Chen, D.M. Frangopol, and X. Ruan, CRC Press, Porto, Portugal.

Hesham, A., Nowak, A., Stallings, M., Chmielewski, J., Stawska, S., Babu, A., Haddadi, F. (2020), *Impact of heavy trucks and permitted overweight loads on highways and bridges now and in the future versus permit fees, truck registration fees, and fuel taxes*, Agreement No. BE695, Final Report, Florida International University.

Heywood, R., Gordon, R., Boully, G. (2000), 'Australia's bridge design load model: Planning for an efficient road transport industry', *Transportation Research Record*, 1696(1), 1–7. doi:10.3141/1696-36.

Highways Agency (2001), *The assessment of highway bridges and structures*, DMRB, 3, Section 4, Part 3, BD 21/01, London.

Highways Agency (2011), *The assessment of highway bridges and structures for the effects of special types general order (STGO) and special order (SO) vehicles*, DMRB, 3, Section 4, Part 19, BD 86/11, London.

Hitchcock, W.A., Salama, T., Zhao, M.H., Callahan, D., Toutanji, H. (2011), *Expanding portable B-WIM technology*, UTCA Project No. 08204.

Hoel, P.G. (1962), *Introduction to mathematical statistics*, 3rd ed., Wiley, New York.

Hong, F., Prozzi, J.A. (2006), 'Estimation of pavement performance deterioration using Bayesian approach', *Journal of Infrastructure Systems*, 12(2), 77–86.

Hoogendoorn, S., Hoogendoorn, R. (2010), 'Calibration of microscopic traffic-flow models using multiple data sources', *Philosophical Transactions of the Royal Society A: Mathematical, Physical and Engineering Sciences*, 368(1928), 4497–4517.

HSBA (2001), *Honshu-Shikoku bridge authority superstructure design standard*, Kobe, Japan.

Huang, D. (2008), 'Dynamic loading of curved steel box girder bridges due to moving vehicles', *Structural Engineering International*, 4, 365–372.

Huang, D.H., Wang, T.L., Shahawy, M. (1993), 'Impact studies of multigirder concrete bridges', *Journal of Structural Engineering*, ASCE, 119(8), 2387–2402.

Hussein, A., González, A. (2012), 'Response of a simply supported beam with a strain rate dependent elasticity modulus when subjected to a moving load', *Bridge and Concrete Research in Ireland*, BCRI, Eds. C. Caprani and A. O'Connor, 6–7 September.

Hwang, E., Nowak, A.S. (1991), 'Simulation of dynamic load for bridges', *Journal of Structural Engineering*, ASCE, 117, 1413–1434. doi: 10.1061/ (ASCE)0733-9445(1991)117:5(1413).

Hwang, E.-S., Kim, D.-Y. (2019), 'Live load model for long span steel cable bridges considering traffic congestion scenarios', *International Journal of Steel Structures*, 19, 1996–2009.

Hwang, E.S., Kim, D.Y., Mok, J.Y. (2013), *Design live load model for long span bridges. 36th International Association for Bridge and Structural Engineering Symposium on Long Span Bridges and Roofs – Development, Design and Implementation*, Kolkata; India.

Iatsko, O. (2018), *WIM-based live load model for bridges*, Doctoral dissertation, Auburn University, Auburn, AL.

Igusa, T., Buonopane, S.G., Ellingwood, B.R. (2002), 'Bayesian analysis of uncertainty for structural engineering applications', *Structural Safety*, 24(2), 165–186.

Iman, R.L., Conover, W.J. (1982), 'A distribution-free approach to inducing rank correlation among input variables', *Communications in Statistics – Simulation and Computation*, 11(3), 311–334.

IRD (2020), *IRD virtual weigh station*, Commercial Vehicle Enforcement, https://www.irdinc.com/pages/its-solutions/virtual-weigh-stations.html (accessed November 2020).

ISO (2015), *ISO 2394:2015 – General principles on reliability for structures*, International Organization for Standardization. doi: 10.1007/s11367-011-0297-3.

Ivy, R., Lin, T., Mitchell, S., Raab, N., Richey, V., Scheffey, C. (1954), 'Live loading for long-span highway bridges', *American Society of Civil Engineers Transactions*, 119, 981–1004.

Jacob, B. (1991), *Methods for the prediction of extreme vehicular loads and load effects on bridges*, Report of Subgroup 8, Eurocode 1, Traffic Loads on Bridges, LCPC Publications, Paris, France, 8.

Jacob, B. (1995), 'European research activity COST 323: Weigh-in-motion of road vehicles', *1st European Conference on Weigh-in-Motion of Road Vehicles*, Zurich, March 8–10.

Jacob, B., Bereni, B.A., Ghosn, M., Sivakumar, B., Kearney, T. (2010), 'US bridge formula (FBF-B) and implications of its possible application in Europe', *Bridge Maintenance, Safety, Management and Life-Cycle Optimization*, Eds. Frangopol, Sause, and Kusko, Taylor & Francis Group, Philadelphia, PA, 2875–2880.

James, G. (2005), 'Analysis of traffic load effects on railway bridges using weigh-in-motion data', *Fourth International Conference on Weigh-In-Motion (ICWIM4)*, Eds. E.J. OBrien, B. Jacob, A. González, and C.-P. Chou, National Taiwan University, 351–361.

Jenkinson, A.F. (1955), 'The frequency distribution of the annual maximum (or minimum) of meteorological elements', *Quarterly Journal of the Royal Meteorological Society*, 81, 158–171.

Jones, M.C., Marron, J.S., Sheather, S.J. (1996), 'A brief survey of bandwidth selection for density estimation', *Journal of the American Statistical Association*, 91(433), 401–407.

JRA (1996), *Japan Road Association specifications for highway bridges. Part 1: Common specifications*, JRA, Tokyo.

Kalin, J., Žnidarič, A., Anžlin, A., Kreslin, M. (2021), 'Measurements of bridge dynamic amplification factor using bridge weigh-in-motion data', *Structure & Infrastructure Engineering*. doi: 10.1080/15732479.2021.1887291.

Kalin, J., Žnidarič, A., Kreslin, M. (2015), *Using weigh-in-motion data to determine bridge dynamic amplification factor*, MATEC Web of Conferences, Dübendorf, 02003–02008.

Kanda, J., Ellingwood, B. (1991), 'Formulation of load factors based on optimum reliability', *Structural Safety*, 9(3), 197–210.

Kesting, A., Treiber, M. (2008a), 'Calibrating car-following models by using trajectory data', *Transportation Research Record: Journal of the Transportation Research Board*, 2088, 148–56.

Kesting, A., Treiber, M. (2008b), 'How reaction time, update time, and adaptation time influence the stability of traffic flow', *Computer-Aided Civil and Infrastructure Engineering*, 23, 125–137.

Kesting, A., Treiber, M., Helbing, D. (2007), 'General lane-changing model MOBIL for car-following models', *Transportation Research Record: Journal of the Transportation Research Board*, 1999(1), 86–94.

Kim, C.W., Kawatani, M., Kwon, Y.R. (2007), 'Impact coefficient of reinforced concrete slab on a steel girder bridge', *Engineering Structures*, 29, 576–590.

Kim, K.-D., Heo, J.-H. (2002), 'Comparative study of flood quantiles estimation by nonparametric models', *Journal of Hydrology*, 260(1), 176–193.

Kim, S., Nowak, A.S. (1997), 'Load distribution and impact factors for I-girder bridges', *Journal of Bridge Engineering*, 2(3), 97–104.

King, L.J. (1984), *Central place theory*, Sage, London.

Kirkegaard, P., Neilsen, S., Enevoldsen, I. (1997), *Heavy vehicles on minor highway bridges - Calculation of dynamic impact factors from selected crossing scenarios*, Department of Building Technology and Structural Engineering, Aalborg University, Aalborg.

Klein, L.A., Mills, M.K., Gibson, D.R.P. (2006), *Traffic detector handbook*, 3rd ed., V.A. MacLean, Federal Highway Administration.

Koniditsiotis, C., Coleman, S., Cai, D. (2012), 'Bringing heavy vehicle on-board mass monitoring to market', *6th International Conference on Weigh-in-Motion*, Eds. B. Jacob, A.-M. McDonnell, F. Schmidt, and C. Wiley, Dallas, TX, 304–318.

Kroese, D.P., Taimre, T., Botev, Z.I. (2011), *Handbook of Monte Carlo methods*, John Wiley & Sons, New York.

Kulicki, J.M., Prucz, Z., Clancy, C.M., Mertz, D., Nowak, A. (2007), *Updating the calibration report for AASHTO LRFD code*, NCHRP 20-7/186, Transport Research Board, Washington, DC.

Kulicki, J.M., Wasser, W.G., Mertz, D.R., Nowak, A.S. (2015), *Bridges for service life beyond 100 years: Service limit state design*, SHRP 2 Renewal Project R19B, Transportation Research Board, Washington, DC.

Kwon, O., Kim, E., Orton, S. (2011), 'Calibration of the live load factor in LRFD design specifications based on state-specific traffic environments', *Journal of Bridge Engineering*, ASCE, 16(6), 812–819.

Laman, J.A., Nowak, A.S. (1996), 'Fatigue-load models for girder bridges', *Journal of Structural Engineering*, 122, 726–733. doi: 10.1061/ (ASCE)0733-9445(1996)122:7(726).

Laman, J.A., Pechar, J.S., Boothby, T.E. (1999), 'Dynamic load allowance for through-truss bridges', *Journal of Bridge Engineering*, 4(4), 231–241.

Leadbetter, M.R., Lindgren, G., Rootzen, H. (1983), *Extremes and related properties of random sequences and processes*, The University of California, CA, Springer-Verlag.

Leahy, C. (2014), *Predicting extreme traffic loading on bridges using weigh-in-motion data*, PhD thesis, University College Dublin, Ireland.

Leahy, C., OBrien, E.J. (2011), *Bridge weigh in motion systems*, Report to the National Roads Authority. Report No. 2011-1, UCD School of Civil, Structural and Environmental Engineering, Dublin, Ireland.

Leahy, C., OBrien, E.J., Enright, B., Hajializadeh, D. (2014), 'A review of the HL-93 bridge traffic load model using an extensive WIM database', *Journal of Bridge Engineering*, ASCE, 20(10), doi: 10.1061/(ASCE)BE.1943-5592.0000729.

Lenner, R., de Wet, D.P.G., Viljoen, C. (2017), 'Traffic characteristics and bridge loading in South Africa', *Journal of the South African Institution of Civil Engineering*, 59(4), 34–46. doi: 10.17159/2309-8775/2017/v59n4a4.

Li, H., Wekezer, J., Kwasniewski, L. (2008), 'Dynamic response of a highway bridge subjected to moving vehicles', *Journal of Bridge Engineering*, 13(5), 439–448.

Li, Y. (2006), *Factors affecting the dynamic interaction of bridges and vehicle loads*, PhD thesis, University College Dublin, Ireland.

Li, Y., OBrien, E.J., González, A. (2006), 'The development of a dynamic amplification estimator for bridges with good road profile', *Journal of Sound and Vibration*, 293(1–2), 125–137.

Liao, C.-F., Chatterjee, I., Davis, G.A. (2015), *Implementation of traffic data quality verification for WIM sites*, Report, Center for Transportation Studies University of Minnesota.

Lima, J.M., Brito, J. (2009), 'Inspection survey of 150 expansion joints in road bridges', *Engineering Structures*, 31, 1077–1084.

Lingappaiah, G. (1984), 'Bayesian prediction regions for the extreme order statistics', *Biometrical Journal*, 26(1), 49–56.

Lipari, A. (2013), *Micro-simulation modelling of traffic loading on long-span bridges*, PhD Thesis, University College Dublin, Ireland.

Lipari, A., Caprani, C.C., OBrien, E.J. (2017a), 'A methodology for calculating congested traffic characteristic loading on long-span bridges using site-specific data', *Computers & Structures*, 190(1), October, 1–12. doi: 10.1016/j.compstruc.2017.04.019.

Lipari, A., Caprani, C.C., OBrien, E.J. (2017b), 'Heavy-vehicle gap control for bridge traffic loading mitigation', *IEEE Intelligent Transport Systems Magazine*, 9(4), 118–131. doi: 10.1109/MITS.2017.2743169.

Ludescher, H., Brühwiler, E. (2009), 'Dynamic amplification of traffic loads on road bridges', *Structural Engineering International*, May, 190–197.

Luskin, D.L., Harrison, R., Walton, C.M., Zhang, Z., Jamieson, J.L.J. (2000), *Alternatives to weight tolerance permits*, FHWA, Research Report 0-4036-1, Austin, TX.

Luskin, D.L., Walton, C.M. (2001), *Synthesis study of the effects of overweight/oversize trucks*, Center for Transportation Research, Bureau of Engineering Research, University of Texas at Austin.

Malevergne, Y., Pisarenko, V., Sornette, D. (2006), 'On the power of generalized extreme value (GEV) and generalized Pareto distribution (GPD) estimators for empirical distributions of log-returns', *On the power of generalized extreme value (GEV) and generalized Pareto distribution (GPD) estimators for empirical distributions of log-returnsApplied Financial Economics*, 16(3), 271–289.

Marková, J. (2013), 'Reliability assessment of traffic load models on road bridges', *IABSE Symposium Report*, 99, 600–605. doi: 10.2749/222137813806481446.

McCall, B., Vodrazka, W.J. (1997), *States' successful practices weigh-in-motion handbook*, Center for Transportation Research and Education, Iowa State University, IA.

McLean, D., Marsh, M.L. (1998), *Dynamic impact factors for bridges*, National Cooperative Highway Research Program (NCHRP) Synthesis 266. Transportation Research Board, National Research Council (US).

Melchers, R.E. (1999), *Structural reliability: Analysis and prediction*, John Wiley & Sons, New York.

Melchers, R.E. (2001), 'Assessment of existing structures – Approaches and research needs', *Journal of Structural Engineering*, 127(4), 406–411.

Melhem, M., Caprani, C.C., Stewart, M.G., Zhang, S. (2020), *Bridge Assessment Beyond the AS 5100 Deterministic Methodology*, AustRoads Report AP-R617-20, Sydney.

Mendez, F.J., Menendez, M., Luceño, A., Losada, I.J. (2007), 'Analyzing monthly extreme sea levels with a time-dependent GEV model', *Journal of Atmospherica and Oceanic Technology*, 24, 894–911.

Menendez, M., Mendez, F.J., Izaguirre, C., Luceño, A., Losada, I.J. (2009), 'The influence of seasonality on estimating return values of significant wave height', *Coastal Engineering*, 56(3), 211–219.

Miao, T.J., Chan, T.H.T. (2002), 'Bridge live load models from WIM data', *Engineering Structures*, 24(8), 1071–1084.

Michaltsos, G.T. (2000), 'Parameters affecting the dynamic response of light (steel) bridges', *Facta Universitatis, Series Mechanics, Automatic Control and Robotics*, 2(10), 1203–1218.

Michaltsos, G.T., Konstantakopoulos, T.G. (2000), 'Dynamic response of a bridge with surface deck irregularities', *Journal of Vibration and Control*, 6, 667–689.

Micu, E.A., Malekjafarian, A., OBrien, E.J., Quilligan, M., McKinstray, R., Angus, E., Lydon, M., Catbas, F.N. (2019), 'Evaluation of the extreme traffic load effects on the Forth Road Bridge using image analysis of traffic data', *Advances in Engineering Software*, 137, 102711. doi: 10.1016/j.advengsoft.2019.102711.

Micu, E.A., OBrien, E.J., Malekjafarian, A., McKinstray, R., Angus, E., Lydon, M., Catbas, N. (2020), 'Identifying critical clusters of traffic-loading events in recurrent congested conditions on a long-span road bridge', *Applied Sciences*, 10(16), 5423. doi: 10.3390/app10165423.

Micu, E.A., OBrien, E.J., Malekjafarian, A., Quilligan, M. (2018), 'Estimation of extreme load effects on long-span bridges using traffic image data', *Baltic Journal of Road and Bridge Engineering*, 13(4). doi: 10.7250/bjrbe.2018-13.427.

Miner, M.A. (1954), 'Cumulative damage in fatigue', *Journal of Applied Mechanics*, 12, A159–A164.

Moon, Y., Lall, U. (1994), 'Kernel quantite function estimator for flood frequency analysis', *Water Resources Research*, 30(11), 3095–3103.

Moses, F. (2001), *Calibration of load factors for LRFR bridge evaluation*, NCHRP Report 454, Transport Research Board, Washington, DC.

Moses, F., Ghosn, M. (1984), 'Application of load spectra to bridge rating', 2nd Bridge Engineering Conference conducted by the Transportation Research Board and the Federal Highway Administration, *Transportation Research Record*, 1(950), https://trid.trb.org/view/210125.

Moyo, P., Brownjohn, J.M., Omenzetter, P. (2003), 'Highway bridge live loading assessment and load carrying estimation using a health monitoring system', *Proceedings of the 3rd International Conference on Current and Future Trends in Bridge Design, Construction and Maintenance*, Eds. B.I.G. Barr et al., Shanghai, September/October, Thomas Telford, 557–564.

MTRPC (1989), General code for design of highway bridges and culverts, Ministry of Transport of the People's Republic of China (MTPRC), Beijing.

Nadarajah, S. (2005), 'Extremes of daily rainfall in west central Florida', *Climatic Change*, 69, 325–342.

Naess, A., Leira, B.J., Batsevych, O. (2009), 'System reliability analysis by enhanced Monte Carlo simulation', *Structural Safety*, 31(5), 349–355.

Nassif, H.H., Nowak, A.S. (1995), 'Dynamic load spectra for girder bridges'. *Transportation Research Record*, 1476, 69–83.

Nassif, H.H., Nowak, A.S. (1996), 'Dynamic load for girder bridges under normal traffic', *Archives of Civil Engineering*, XLII(4), 381–400.

Navin, F.P.D., Zidek, J.V., Fisk, C., Buckland, P.G. (1976), 'Design traffic loads on the lions' gate bridge', *Transportation Research Record*, 607.

Newell, G.F. (1961), 'Nonlinear effects in the dynamics of car following', *Operations Research*, 9(2), 209–229.

Newmark, M.M. (1948), 'Design of I-beam bridges', 'Hwy. Bridge Floor Symp.', *Journal of Structural Division*, ASCE, 24(2), 141–161.

Nogaj, M., Parey, S., Dacunha-Castelle, D. (2007), 'Non-stationary extreme models and a climatic application', *Nonlinear Processes in Geophysics*, 14, 305–316.

Norman, O.K., Hopkins, R.C. (1952), *Weighing vehicles in motion*. Bulletin 50, Highway Research Board, Washington, DC.

Nowak, A. (1993), 'Live load model for highway bridges', *Structural Safety*, 13(1–2), 53–66. doi: 10.1016/0167-4730(93)90048-6.

Nowak, A. (1999), *Calibration of LRFD bridge design code*, NCHRP Report 368, Transport Research Board, Washington, DC, https://trid.trb.org/view.aspx ?id=492076 (accessed 20 September 2016).

Nowak, A., Rakoczy, P. (2012), 'WIM based simulation model of site specific live load effect on the bridges', *6th International Conference on Weigh-In-Motion*, Eds. B. Jacob, A.-M. McDonnell, F. Schmidt, and C. Wiley, Dallas, TX, 352–358.

Nowak, A., Szerszen, M.M. (1998), 'Bridge load and resistance models', *Engineering Structures*, 20(11), 985–990.

Nowak, A.S. (1989), 'Probabilistic basis for bridge design codes', 5th International Conference on Structural Safety and Reliability, ICOSSAR '89, San Francisco, CA, Eds. A.H.-S. Ang, M. Shinozuka and G.I. Schuëller, ASCE, New York, 2019–2026.

Nowak, A.S. (1994), 'Load model for bridge design code', *Canadian Journal of Civil Engineering*, 21(1), 36–49.

Nowak, A.S. (1995), Reply to: 'Load model for bridge design code', *Canadian Journal of Civil Engineering*, 22(2): 293–293.

Nowak, A.S., Collins, K.R. (2013), *Reliability of structures*, 2nd ed., CRC Press, Boca Raton, FL.

Nowak, A.S., Grouni, H.N. (1994), 'Calibration of the Ontario highway bridge design code 1991 edition', *Canadian Journal of Civil Engineering*, 21(1), 25–35.

Nowak, A.S., Hong, Y.K. (1991), 'Bridge live load models', *Journal of Structural Engineering*, ASCE, 117(9), 2757–2767.

Nowak, A.S., Iatsko, O. (2017), 'Revised load and resistance factors for the AASHTO LRFD bridge design specifications', *PCI Journal*, 3, 46–58.

Nowak, A.S., Lind, N.C. (1979), 'Practical bridge code calibration', *ASHRAE Journal*, 105, American Society of Heating, Refrigerating and Air-Conditioning Engineers.

Nowak, A.S., Lutomirska, M., Sheikh Ibrahim, F.I. (2010), 'The development of live load for long span bridges', *Bridge Structures*, 6, 73–79.

Nowak, A.S., Nassif, H., DeFrain, L. (1993), 'Effect of truck loads on bridges', *Journal of Transportation Engineering*, ASCE, 119(6), 853–867.

Nowak, A.S., Tharmabala, T. (1988), 'Bridge reliability evaluation using load tests', *Journal of Structural Engineering*, 114, 2268–2279. doi: 10.1061/(ASCE)0733-9445(1988)114:10(2268).

OBrien, E.J., Bordallo-Ruiz, A., Enright, B. (2014), 'Lifetime maximum load effects on short-span bridges subject to growing traffic volumes', *Structural Safety*, 50, 113–122. doi: 10.1016/j.strusafe.2014.05.005.

OBrien, E.J., Cantero, D., Enright, B., González, A. (2010a), 'Characteristic dynamic increment for extreme traffic loading events on short and medium span highway bridges', *Engineering Structures*, 32(12), 3827–3835.

OBrien, E.J., Caprani, C.C. (2005), 'Headway modelling for traffic load assessment of short to medium span bridges', *Structural Engineer*, 83(16), 33–36.

OBrien, E., Caprani, C.C., Blacoe, S., Guo, D., Malekjafarian, A. (2018), 'Detection of vehicle wheels from images using a pseudo-wavelet filter for analysis of congested traffic', *IET Image Processing*, 12(12), 2222–2228. doi: 10.1049/iet-ipr.2018.5369.

OBrien, E.J., Caprani, C.C., O'Connell, G.J. (2006), 'Bridge assessment loading: A comparison of West and Central/East Europe', *Bridge Structures*, 2, 25–33. doi: 10.1080/15732480600578451.

OBrien, E.J., Enright, B. (2011), 'Modeling same direction two lane traffic for bridge loading', *Structural Safety*, 33(4–5), 296–304. doi: 10.1016/j.strusafe.2011.04.004.

OBrien, E.J., Enright, B. (2012), 'Using weigh-in-motion data to determine aggressiveness of traffic for bridge loading', *Journal of Bridge Engineering*, ASCE, 18(3), 232–239. doi: 10.1061/(ASCE)BE.1943-5592.0000368.

OBrien, E.J., Enright, B., Getachew, A. (2010b), 'Importance of the tail in truck weight modelling for bridge assessment', *Journal of Bridge Engineering*, ASCE, 15(2), 210–213.

OBrien, E.J., Enright, B., Leahy, C. (2013), 'The effect of truck permitting policy on US bridge loading', *11th International Conference on Structural Safety & Reliability*, Eds. G. Geodatis, B. R. Ellingwood, and D. M. Frangopol, New York.

OBrien, E.J., González, A., Dowling, J., Žnidarič, A. (2013a), 'Direct measurement of dynamics in road bridges using a bridge-weigh-in-motion system', *The Baltic Journal of Road and Bridge Engineering*, 8(4), 263–270. doi: 10.3846/bjrbe.2013.34.

OBrien, E.J., Hayrapetova, A., Walsh, C. (2012a), 'The use of micro-simulation for congested traffic load modelling of medium- and long-span bridges', *Structure and Infrastructure Engineering*, 8, 269–76.

OBrien, E.J., Leahy, C., Enright, B., Caprani, C.C. (2016), 'Validation of Scenario Modelling for Bridge Loading', *The Baltic Journal of Road and Bridge Engineering*, 11(3), Sep., 233–241.

OBrien, E.J., Lipari, A., Caprani, C.C. (2015a), 'Micro-simulation of single-lane traffic to identify critical loading conditions for long-span bridges', *Engineering Structures*, 94, 137–148. doi: 10.1016/j.engstruct.2015.02.019.

OBrien, E.J., O'Connor, A., Arrigan, J.E. (2012b), 'Procedures for calibrating Eurocode traffic Load Model 1 for national conditions', *Proceedings of the 6th International Conference on Bridge Maintenance, Safety and Management*, Eds. F. Biondini and D.M. Frangopol, CRC Press, Stresa, Italy, 2597–2603.

OBrien, E.J., Rattigan, P., González, A., Dowling, J., Žnidarič, A. (2009a), 'Characteristic dynamic traffic load effects in bridges', *Engineering Structures*, 31(7), 1607–1612.

OBrien, E.J., Rowley, C.W., González, A., Green, M.F. (2009b), 'A regularised solution to the bridge weigh in motion equations', *International Journal of Heavy Vehicle Systems*, 16(3), 310–327. doi: 10.1504/IJHVS.2009.027135.

OBrien, E.J., Schmidt, F., Hajializadeh, D., Zhou, X.-Y., Enright, B., Caprani, C.C., Wilson, S., Sheils, E. (2015b), 'A review of probabilistic methods of assessment of load effects in bridges', *Structural Safety*, 53, March, 44–56. doi: 10.1016/j.strusafe.2015.01.002.

OBrien, E., Žnidarič, A., Ojio, T. (2008), 'Bridge weigh-in-motion – Latest developments and applications world wide', *Proceedings of the International Conference on Heavy Vehicles*, Eds. B. Jacob, E.J. OBrien, P. Nordengen, A. O'Connor, and M. Bouteldja, ISTE/Hermes, London, 19–22.

O'Connor, A., Eichinger, E. (2007), 'Site-specific traffic load modelling for bridge assessment', *Proceedings of the Institution of Civil Engineers. Bridge Engineering*, 160(4), 185–194.

O'Connor, A., Enevoldsen, I. (2009), 'Probability-based assessment of highway bridges according to the new Danish guideline', *Structure and Infrastructure Engineering*, 5(2), 157–168.

O'Connor, A., Jacob, B., OBrien, E.J., Prat, M. (2001), 'Report of current studies performed on normal load model of EC1: Part 2, traffic loads on bridges', *Revue française de génie civil*, 5(4), 411–433.

O'Connor, A., OBrien, E.J. (2005), 'Traffic load modelling and factors influencing the accuracy of predicted extremes', *Canadian Journal of Civil Engineering*, 32(1), 270–278. doi: 10.1139/l04-092.

O'Connor, A., Pedersen, C., Gustavsson, L., Enevoldsen, I. (2009), 'Probability-based assessment and optimised maintenance management of a large riveted truss railway bridge', *Structural Engineering International*, 4, 375–382.

Orosz, G., Wilson, R.E., Stépán, G. (2010), 'Traffic jams: Dynamics and control', *Philosophical Transactions of the Royal Society A*, 368(1928), 4455–4479.

O'Sullivan, A. (2009), *Urban economics*, McGraw-Hill/Irwin, Boston, MA.

Papagiannakis, T., Senn, K., Huang, H. (1995), *On-site evaluation and calibration procedures for weigh-in-motion systems*, American Association of State Highway and Transportation Officials, Washington, DC.

Pape, T., Kotze, R., Ngo, H. (2014), 'Dynamic bridge-vehicle interactions', *9th AustRoads Bridge Conference*, Sydney, Retreived from https://www.online-publications.austroads.com.au/items/ABC-SAS301-14

Paxton, R.A. (1977), 'Menai Bridge (1818–1826) and its influence on suspension bridge development', *Transactions of the Newcomen Society*, 49, 87–110.

Pelphrey, J., Higgins, C., Sivakumar, B., Groff, R.L., Hartman, B.H., Charbonneau, J.P., Johnson, B.V. (2008), 'State-specific LRFR live load factors using weigh-in-motion data', *Journal of Bridge Engineering*, 13(4), 339–350.

PENN Bridge Company (1886), *Iron highway bridges: As built by the Penn Bridge Company*, J. W. Shipman, Beaver Falls, PA.

PennDOT (2012), *Design manual – Part 4, structures*, Department of Transportation, Harrisburg, PA.

Pérez Sifre, S., Lenner, R. (2019), 'Bridge assessment reduction factors based on Monte Carlo routine with copulas', *Engineering Structures*, 198(August), 109530. doi: 10.1016/j.engstruct.2019.109530.

Phoenix Bridge Company (1885), *Album of designs of the Phoenix Bridge Company*, J.B. Lippincott & Co, Philadelphia, PA.

Pickands, J. (1975), 'Statistical inference using extreme order statistics', *Annals of Statistics*, 3, 119–131.

Prat, M. (2001), Traffic load models for bridge design: recent developments and research, *Progress in Structural Engineering and Materials*, 3, pp. 326–334.

Press, W.H., Teukolsky, S.A., Vetterling, W.T., Flannery, B.P. (2007), *Numerical recipes: The art of scientific computing*, 3rd ed., Cambridge University Press, Cambridge, UK.

Punzo, V., Simonelli, F. (2005), 'Analysis and comparison of microscopic traffic flow models with real traffic microscopic data', *Transportation Research Record: Journal of the Transportation Research Board*, 1934(1), 53–63.

Qu, T., Lee, C.E., Huang, L. (1997), *Traffic-load forecasting using weigh-in-motion data*, Research Report 987–6, Center For Transportation Research, Bureau of Engineering Research, University of Texas at Austin, March.

Quinley, R. (2010), *WIM data analyst's manual*, Federal Highway Administration Report No. FHWA-IF-10-018, Washington, DC.

Rackwitz, R. (2000), 'Optimization – The basis of code-making and reliability verification', *Structural Safety*, 22(1), 27–60. doi: 10.1016/S0167-4730(99)00 037-5.

Ramachandran, A.N., Taylor, K.L., Stone, J.R., Sajjadi, S.S. (2011), 'NCDOT quality control methods for weigh-in-motion data', *Public Works Management & Policy*, 16(1), 3–19. doi: 10.1177/1087724X10383583.

Rattigan, P.H. (2007), *The assessment of bridge traffic loading allowing for vehicle-bridge dynamic interaction*, PhD Dissertation, University College Dublin, Ireland.

Reale, T., O'Connor, A. (2012), 'Cross entropy as an optimisation method for bridge condition transition probability determination', *ASCE Journal of Transportation Engineering*, 138(6), 741 –751.

Renard, B., Lang, M., Bois, P. (2006), 'Statistical analysis of extreme events in a non-stationary context via a Bayesian framework: Case study with peak-over-threshold data', *Stochastic Environmental Research and Risk Assessment*, 21, 97–112.

Ribereau, P., Guillou, A., Naveau, P. (2008), 'Estimating return levels from maxima of non-stationary random sequences using the generalized PWM method', *Nonlinear Processes in Geophysics*, 15, 1033–1039.

Rice, S. (1945), 'The mathematical analysis of random noise', *Bell System Technical Journal*, 24, 24–156.

Ricketts, N.J., Page, J. (1997), *Traffic data for highway bridge loading*, Transport Research Laboratory, Crowthorne.

Rocco, M. (2010), *Extreme value theory for finance: A survey*, Statistics and Informatics Working Paper No. 3, Department of Mathematics, University of Bergamo, Italy.

Rogers, M. (2008), *Highway engineering*, 2nd ed., Blackwell, Oxford.

Ruan, X., Shi, X., Ying, T. (2010), 'Analysis of highway vehicle load in China based on WIM data', *IABSE Symposium Report*, 97, 55–62.

Ruan, X., Zhou, J., Caprani, C.C. (2016), 'Safety assessment of the antisliding between the main cable and middle saddle of a three-pylon suspension bridge consider traffic load modeling', *Journal of Bridge Engineering*, ASCE, 21(10), 04016069.

Ruan, X., Zhou, J., Shi, X., Caprani, C.C. (2017a), 'A site-specific traffic load model for long-span multi-pylon cable-stayed bridges', *Structure and Infrastructure Engineering*, 13(4), 494–504. doi: 10.1080/15732479.2016.1164724.

Ruan, X., Zhou, J., Tu, H., Jin, Z., Shi, X. (2017b), 'An improved cellular automation with axis information for microscopic traffic simulation', *Transportation Research Part C*, 78, 63–77.

Ruan, X., Zhou, K.P. (2014), 'Vehicle characteristics and load effects of four-lane highway', *7th International Conference of Bridge Maintenance, Safety and Management*, Shanghai, Taylor and Francis.

Ruan, X., Zhou, X.Y., Guo, J. (2012), 'Extreme value extrapolation for bridge vehicle load effect based on synthetic vehicle flow', *Journal of Tongji University: Natural Science*, 40, 1458–1485 (In Chinese).

Rubinstein, R. (1999), 'The cross-entropy method for combinatorial and continuous optimization', *Methodology and Computing in Applied Probability*, 1(2), 127–190.

Rubinstein, R., Kroese, D. (2004), *The cross-entropy method*, Springer, New York.

SAMARIS (2006), *Guideline for the optimal assessment of highway structures*, Sustainable and Advanced MAterials for Road InfraStructure (SAMARIS), Deliverable SAM-GE-D30, EU 6th Framework Report, http://samaris.zag.si/.

Sanders, W.W., Elleby, H.A. (1970), *Distribution of wheel loads on highway bridges*, National Cooperative Highway Research Program, National Research Council (USA), Highway Research Board, Washington, DC.

Sayed, S.M., Sunna, H.N., Moore, P.R. (2020), 'Truck platooning impact on bridge preservation', *Journal of Performance of Constructed Facilities*, ASCE, 34(3), 04020029.

Schulman, J.F. (2003), *Heavy truck weight and dimension limits in Canada*, Railway Association of Canada.

Scott, D.W. (1992), *Multivariate density estimation: Theory, practice, and visualization*, John Wiley & Sons, New York.

Seaman, H.B. (1912), 'Specifications for the design of bridges and subways', *Proceedings of the American Society of Civil Engineers*, 37, 1261–1300.

Sedlacek, G., Merzenich, G., Paschen, M., Bruls, A., Sanpaolesi, L., Croce, P., Calgaro, J.A., Pratt, M., Jacob, L.M., Boer, V.A. and Hanswille, G. (2008), *Background document to EN 1991-Part 2-Traffic loads for road bridges – and consequences for the design*, JRC Scientific and Technical Reports, European Commission Joint Research Centre.

SHRP2 (2015), Bridges for Service Life Beyond 100 Years: Service Limit State Design, SHRP2 Report S2-R19B-RW-1, Transportation Research Board, Washington DC.

Silverman, B.W. (1986), *Density estimation for statistics and data analysis*, Chapman & Hall, London.

Sinha, S.K., Sloan, J.A. (1988), 'Bayes estimation of the parameters and reliability function of the 3-parameter Weibull distribution', *IEEE Transactions on Reliability*, 37(4), 364–369.

Sivakumar, B. (2007), *Legal truck loads and AASHTO legal loads for posting*, NCHRP Report 575, Transportation Research Board.

Sivakumar, B., Ghosn, M., Moses, F. (2011), *Protocols for collecting and using traffic data in bridge design*, NCHRP Report 683, Transport Research Board, Washington, DC. doi: 10.17226/14521.

Sivakumar, B., Moses, F., Fu, G., Ghosn, M. (2007), *Legal truck loads and AASHTO legal loads for posting*, NCHRP Report 575.

Soriano, M., Casas, J.R., Ghosn, M. (2017), 'Simplified probabilistic model for maximum traffic load from weigh-in-motion data', *Structure and Infrastructure Engineering*, 13(4), 454–467. doi: 10.1080/15732479.2016.1164728.

Southgate, H.F. (1990), *Estimation of equivalent axleloads using data collected by automated vehicle classification and weigh-in-motion equipment*, Research Report 485, Kentucky Transportation Center Research.

Sparmann, U. (1979), 'The importance of lane-changing on motorways', *Traffic Engineering and Control*, 20(6), 320–323.

Srinivas, S., Menon, D., Meher Prasad, A. (2006), 'Multivariate simulation and multimodal dependence modeling of vehicle axle weights with copulas', *Journal of Transportation Engineering*, ASCE, 132(12), 945–955.

Standards Australia (2017a), *AS 5100.2-2017: Bridge design – Design loads*, Sydney.

Standards Australia (2017b), *AS 5100.7-2017: Bridge design – Bridge assessment*, Sydney.

Stathopolous, A., Karlaftis, M. (2001), 'Temporal and spatial variations of real-time traffic data in urban areas', *Journal of the Transportation Research Board*, TRB, 1768, 135–140.

Stawska, S., Chmielewski, J., Bacharz, M., Bacharz, K., Nowak, A. (2021), 'Comparative Accuracy Analysis of Truck Weight Measurement Techniques', Applied Sciences, 11(745), doi: 10.3390/app11020745.

Steenbergen, R.D.J.M., Vrouwenvelder, A.C.W.M. (2010), 'Safety philosophy for existing structures and partial factors for traffic loads on bridges', *Heron*, 55(2), 123–140.

Stefanakos, C.N., Athanassoulis, G.A. (2006), 'Extreme value predictions based on nonstationary time series of wave data', *Environmetrics*, 17, 25–46.

Stephens, J., Carson, J., Hult, D.A., Bisom, D. (2003), 'Preservation of infrastructure by using weigh-in-motion coordinated weight enforcement', *Transportation Research Record: Journal of the Transportation Research Board*, 1855(1), 143–150.

Taylor, H.P.J., Clark, L., Banks, C.C. (1990), 'Y-beam: A replacement for the M-beam in beam and slab bridge decks', *Structural Engineer*, 68(23/4), 459–465.

Tomer, E., Safonov, L., Havlin, S. (2000), 'Presence of many stable nonhomogeneous states in an inertial car-following model', *Physical Review Letters*, 84(2), 382–385.

Treacy, M., Brühwiler, E. (2012), 'Fatigue loading estimation for road bridges using long term WIM monitoring', *Advances in Safety, Reliability and Risk Management, Proceedings of the European Safety and Reliability Annual Conference (ESREL 2011)*, Eds. C. Berenguer, A. Grall, and C.G. Soares, Taylor & Francis Group, Troyes, France, 1870–1875.

Treiber, M., Hennecke, A., Helbing, D. (2000a), 'Congested traffic states in empirical observations and microscopic simulations', *Physical Review E*, 62(2), 1805–1824.

Treiber, M., Hennecke, A., Helbing, D. (2000b), 'Microscopic simulation of congested traffic', *Traffic and Granular Flow '99*, Eds. D. Helbing, H.J. Herrmann, M. Schreckenberg, and D.E. Wolf, Springer, Berlin, 365–376.

Treiber, M., Kanagaraj, V. (2015), 'Comparing numerical integration schemes for time-continuous car-following models', *Physics A: Statistical Mechanics and its Application*, 419C, 183–95.

Turochy, R.E., Timm, D.H., Mai, D. (2015), *Development of Alabama traffic factors for use in mechanistic-empirical pavement design*. No. FHWA/ALDOT 930-793. Auburn University. Highway Research Center.

USDA (1924), *Standard specifications for steel highway bridges: Adopted by the American Association of State Highway Officials and as approved by the Secretary of Agriculture for use in connection with federal-aid road work*, U.S. Department of Agriculture, Washington, DC.

USDOT (2000), *Comprehensive truck size and weight study*, U.S. Department of Transportation, FHWA-PL-00-029.

van der Spuy, P., Lenner, R. (2019), 'Towards a new bridge live load model for South Africa', *Structural Engineering International*, 29(2), 292–298. doi: 10.1080/10168664.2018.1561168.

van der Spuy, P., Lenner, R., de Wet, T., Caprani, C.C. (2019), 'Multiple lane reduction factors based on multiple lane weigh in motion data', *Structures*, 20(April), 543–549. doi: 10.1016/j.istruc.2019.06.001.

Van Dorp, J.R., Mazzuchi, T.A. (2005), 'A general Bayes Weibull inference model for accelerated life testing', *Reliability Engineering & System Safety*, 90(2), 140–147.

Vandervalk-Ostrander, A. (2009), *AASHTO guidelines for traffic data programs*, 2nd ed., American Association of State Highway and Transportation Officials, Washington, DC.

Von Mises, R. (1936), 'La distribution de la plus grande de n valeurs, Revue Mathématique De l'Union Interbalcanique', 1, 141–160. Reproduced in Selected Papers of Richard von Mises, *American Mathematical Society*, 2, 271–294, Rhode Island, 1964.

Vrouwenvelder, A.C.W.M., Waarts, P.H. (1993), 'Traffic loads on bridges', *Structural Engineering International*, 3/93, 169–177.

Waddell, J.A.L. (1886), *The designing of ordinary iron highway bridges*, John Wiley & Sons, New York.

Walker, D., Cebon, D. (2012), 'The metamorphosis of LTPP traffic data', *6th International Conference on Weigh-in-Motion*, Eds. B. Jacob, A.-M. McDonnell, F. Schmidt, and C. Wiley, Dallas, TX, 242–249.

Walker, D., Selezneva, O., Wolf, D.J. (2012), 'Findings from LTPP SPS WIM systems validation study', *6th International Conference on Weigh-in-Motion*, Eds. B. Jacob, A.-M. McDonnell, F. Schmidt, and C. Wiley, Dallas, TX, 214–221.

Walsh, B.J., González, A. (2009), 'Assessment of the condition of a beam using a static loading test', *Key Engineering Materials*, 413, 269–276.

Wang, J., Chaudhury, A., Rao, H.R. (2008), 'A value-at-risk approach to information security investment', *Information Systems Research*, 19(1), 106–120.

Wang, Y., Zongyu, G., Wang, Z., Yang, Y. (2014), 'A case study of traffic load for long-span suspension bridges', *Structural Engineering International*, 24, 352–360.

White, R., Song, J., Haas, C., Middleton, D. (2006), 'Evaluation of quartz piezo-electric weigh-in-motion sensors', *Transportation Research Record: Journal of the Transportation Research Board*, 1945, 109–117. doi: 10.3141/1945-13.

Yarnold, M.T., Wediner, J.S. (2019), 'Truck platoon impacts on steel girder bridges', *Journal of Bridge Engineering*, ASCE, 24(7), (July) 06019003.

Yousif, S., Hunt, J. (1995), 'Modelling lane utilisation on British dual-carriageway roads: Effects on lane-changing', *Traffic Engineering & Control*, 36(12), 680–687.

Zhao, J., Tabatabai, H. (2012), 'Evaluation of a permit vehicle model using weigh-in-motion truck records', *Journal of Bridge Engineering*, 17, 389–392. doi: 10.1061/(ASCE)BE.1943-5592.0000250.

Zhou, J., Liu, Y., Yi, J. (2020), 'Effect of uneven multilane truck loading of multi-girder bridges on component reliability', *Structural Concrete*. doi: 10.1002/suco.201900475.

Zhou, J., Ruan, X., Shi, X., Caprani, C. (2018a), 'An efficient approach for traffic load modelling of long span bridges', *Structure & Infrastructure Engineering*, 15(5), 569–581. doi: 10.1080/15732479.2018.1555264.

Zhou, J., Shi, X., Caprani, C.C., Ruan, X. (2018b), 'Multi-lane factor for bridge traffic load from extreme events of coincident lane load effects', *Structural Safety*, 72, 17–29. doi: 10.1016/j.strusafe.2017.12.002.

Zhou, X.Y. (2013), *Statistical analysis of traffic loads and traffic load effects on bridges*, PhD thesis, Université Paris Est.

Zhou, X.Y., Schmidt, F., Jacob, B. (2012), 'Extrapolation of traffic data for development of traffic load models: Assessment of methods used during background works of the eurocode', *Proceedings of the Sixth International Conference on Bridge Maintenance, Safety and Management*, 1503–1509.

Zhu, X.Q., Law, S.S. (2002), 'Dynamic load on continuous multi-lane bridge deck from moving vehicles', *Journal of Sound and Vibration*, 251(4), 697–716.

Žnidarič, A. (2015), *Heavy-duty vehicle weight restrictions in the EU 28*, Enforcement and Compliance Technologies, 23rd ACEA Scientific Advisory Group Report, European Automobile Manufacturers Association.

Žnidarič, A., Kalin, J., Kreslin, M. (2017), 'Improved accuracy and robustness of bridge weigh-in-motion systems', *Structure & Infrastructure Engineering*, 14(4), 412–424. doi: 10.1080/15732479.2017.1406958.

Žnidarič, A., Lavrič, I., Kalin, J. (2008), 'Measurements of bridge dynamics with a bridge weigh-in-motion system', *5th International Conference on Weigh-in-Motion (ICWIM5)*, Eds. B. Jacob, E.J. OBrien, A. O'Connor, M. Bouteldja, ISTE/Hermes, London, 485–498.

Zokaie, T., Imbsen, R., Osterkamp, T. (1991), 'Distribution of wheel loads on highway bridges', *Transportation Research Record*, 1290, 119–126.

Index